"DISCARD"

PHOTONICS
The New Science of Light

PHOTONICS
The New Science of Light

Valerie Burkig

Drawings by Mike Murphy

ENSLOW PUBLISHERS, INC.

Bloy St. & Ramsey Ave. P.O. Box 38
Box 777   Aldershot
Hillside, N.J. 07205   Hants GU12 6BP
U.S.A.   U.K.

Copyright © 1986 by Valerie C. Burkig

All rights reserved.

No part of this book may be reproduced by any means without the written permission of the publisher.

Library of Congress Cataloging in Publication Data

Burkig, Valerie C.
    Photonics, the new science of light.

   Bibliography: p.
   Includes index.
   Summary: Explains the new area of optic technology dealing with making, using, and manipulating photons of light. It includes lasers, fiber optics, holography, solar cells, and other light-source processing of information.
    1. Photonics—Juvenile literature. [1. Photonics. 2. Optics] I. Title.
TA1521.B87   1986   621.36   85-20715
ISBN 0-89490-107-9

Printed in the United States of America

10 9 8 7 6 5 4 3

Illustration Credits
Courtesy of Bell Laboratories, pp. 70,72,77; Courtesy of Coherent Inc., Palo Alto, California, pp. 63,64; Reprinted Courtesy of Eastman Kodak Company, pp. 101,102,103; Courtesy of Hughes Aircraft Company, pp. 55,60,67,106; Courtesy of Lawrence Livermore National Laboratory, p. 69; Courtesy of MMT Observatory, p. 108; Scripps Institution of Oceanography, p. 92; TRW Photo, pp. 93, 94.

to my husband, Jack

ACKNOWLEDGMENTS

I would like to express my thanks to Dr. Ralph Wuerker at TRW, who supplied photographs for the holography chapter and made valuable suggestions about its content. I'm grateful to Professors Stanley S. Ballard and James R. Brookeman of the physics department at the University of Florida for their critical reviews of the manuscript. My thanks also to Martha Murphy, who read the entire work and suggested many changes to improve clarity of expression. And, finally, I thank Jack Burkig for being a source of help and encouragement throughout.

Contents

Foreword 11

Introduction 13

1 What Light Is 17

2 What Light Does 31

3 Coherent Light: The Laser 47

4 Laser Light in Use 59

5 Light in Fibers 71

6 Reconstructed Light: Holography 81

7 Invisible Light: Infrared 97

8 The Future of Optics 111

Glossary 117

Further Reading 123

Index 125

FOREWORD

For centuries the field of optics existed as one of the stable subdivisions of classical physics. The transition to modern optics began in the early 1900s with the development of the quantum theory, which introduced the photon as the particle of light. This theory helped us understand the interaction of light and matter.

However, the real revolution in the study of optics began in 1960 with the development of a unique source of light—the laser. Since that time, optics has developed into a vibrant field of rapidly expanding horizons. For example, lasers are now widely used in medicine, engineering, and industry. And in the field of communications, messages carried by photons in thin optical fibers are replacing those carried by electrons flowing in copper wires.

This book gives an excellent overview of many exciting aspects of modern optics. The explanations of basics and new developments in this field are clear and accurate without being overly technical. The author's wise selection of diagrams and photographs will be especially helpful to the reader.

I believe that just a few hours spent with this book will be most rewarding. You will become familiar with many topics of increasing importance in this high-technology-oriented world. I recommend *Photonics: The New Science of Light* to students, to teachers, and to others who want to broaden their understanding of optics.

Stanley S. Ballard

Distinguished Service Professor
Emeritus of Physics, University of Florida
Former President, Optical Society of America; Member of its Committee on Education

Introduction

Photonics. There's a strange word. It sounds something like electronics and makes you think of radio and television, calculators and computers. Indeed, some parts of photonics are similar to electronics, but photonics uses units of light, called photons, along with electrons. Photonics includes the familiar field of optics but is much more. You could call it the new optics.

The science of optics has been around for hundreds of years. People have long known about the nature of light, its speed, and the way it reflects and refracts and forms the colors of the rainbow. Optics technology has been applied in making better and better devices: eyeglasses, cameras, microscopes, and telescopes. Optical science has been used to study materials by the way they emit, absorb, or reflect light. In combination with electronics, optics has produced the marvel of television.

Though all these improvements are useful and we enjoy the devices, over the years very little was added to the fundamental theory of optics. Students began to have a rather "ho-hum" feeling about the study of optics.

Then, in 1960, the laser was developed and the optics field exploded. Lasers are now applied in medical research and delicate surgery, in manufacturing, and in solving crimes. They have applications on the battlefield and there are plans to put laser weapons in space. Lasers are used in the supermarket, the chemistry laboratory, and in making those marvelous three-dimensional pictures called holograms.

The fastest developing area for lasers is in communications. Tiny lasers, combined with hairlike optical fibers, have already begun sending telephone messages between cities. Rather than electrons in metal wires to carry conversations, there are photons of light traveling in tiny, clear glass strands. Photons do the job better and more cheaply.

There are even optical processors somewhat similar to computers. While the optical kind are not likely to soon replace digital electronic computers, there are some jobs they do much better. They are very good at improving the quality of satellite images and are useful for missile guidance and for robot vision.

Also in recent years, astronomers have been making exciting discoveries about stars in our Milky Way galaxy and about galaxies billions of light-years away, out toward the edge of the universe. Astronomy is no longer restricted to visible light, but uses infrared, ultraviolet, and X-ray emissions from space. When the Space Telescope is in orbit, there will be even better data. Though visible light passes through the atmosphere, most of the other wavelengths are largely absorbed. A telescope above the atmosphere can collect them full strength. These stronger signals will give scientists a clearer understanding of the universe. The development of new materials and techniques for working with the other wavelengths will also aid scientists.

All these things now can be grouped under the name *photonics*. Photonics is the technology of making and using photons of light, moving them around, amplifying them, and detecting them. Photonics includes lasers and other light sources, fiber optics, optical instruments, and related electronics. It covers

Introduction

not only visible light but infrared and ultraviolet photons. Photonics includes solar cells and making electricity from light. The term applies to using light, instead of electronics, to process information.

We will look at these applications of photonics in the following chapters, but first let's review the "old" optics.

1
What Light Is

Light is a necessity. Without light from the sun, the earth would be a dark planet with no plants, no animals, no people. We wouldn't exist.

Eyesight is important, too. Life is much more difficult and dangerous for people who can't see. Most forms of animal life have some sort of vision.

Since light and vision are so important to people, efforts to understand them probably date to well before historical records. We do know that about 500 B.C. the Greeks believed they could see because something went out from their eyes and touched objects. You still hear people use the expression, "piercing the darkness." Today that seems a strange notion.

It wasn't until the 1600s that scientists agreed seeing depends on something coming from the object to the eye. They did not, however, agree on the nature of that something.

The scientific genius Sir Isaac Newton (1642-1727) believed that light consists of many little particles. He gave an explanation of reflection (light reflects, or bounces off, a mirror) in terms of particles. He also explained refraction (a magnifying glass refracts, or bends, light to a focus) in terms of particles.

There were others who argued instead that light had the properties of waves. The Dutch philosopher Christian Huygens (1629-1695) gave an explanation of reflection and refraction in terms of waves. Though wave theory fit the observations better, Newton was so respected that few people accepted Huygens' ideas.

Many years later, around 1800, Thomas Young (1773-1829) and others finally proved that whatever Newton might have said, light behaved as waves. Young did that by demonstrating the interference of light. Interference is a property of waves.

Waves

To think of a one-dimensional wave, imagine a piece of clothesline with one end fixed. Give the free end of the line a quick up-and-down jerk. As each little segment pulls on the next section, a wave moves along the rope, forming crests (peaks) and troughs (valleys).

If there is a dot painted on the rope, you see that dot go up and down as the wave passes. Note that each section of rope moves vertically. It is the disturbance that goes forward, not the rope.

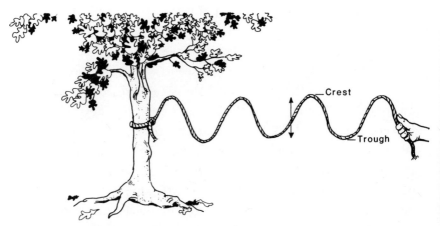

A one-dimensional wave in a rope—each bit of rope moves up and down as the wave progresses.

What Light Is

To see wave motion in two dimensions, drop a stone into a smooth pond—or watch a fish jump. Either one starts the surface water moving up and down, forming crests and troughs. Circular waves progress outward from the original disturbance. Those two-dimensional waves are fairly small, a few inches apart, but easy to see on smooth water. The distance between one crest and the next is called the wavelength. The rate at which each wave goes up and down is its frequency (so many times a second).

If you drop two stones into the pond a short distance from each other, each stone sets up a circular wave traveling outward. When the circles meet, they overlap. At some spots two crests come together. Both waves make the water rise, so it goes higher than if there were just one wave. At other spots the crest of one wave meets the trough of the other. The first will try to make the water go up while the second tries to make it go down. If they are the same height, they neutralize each other so the water remains flat.

That overlapping is called interference. Actually it is not easy to detect the interference pattern in a pond. But it can be shown in the laboratory with two tuning forks, vibrating at the same frequency, immersed in a shallow dish of liquid. That arrangement gives the characteristic pattern seen in the drawing.

The interference pattern made by two equal sources vibrating close together in a liquid. The bright spokes are the high areas where crests add; the dark ones are the flat areas where crests of one meet troughs of the other.

Young's Experiment

Thomas Young produced this pattern with light. He took a plate with a pinhole in it and set it up in front of a light. That gave him, in effect, a point source. He reasoned that a wave would spread out from that hole like water waves from stones.

The statement that light travels in straight lines is not exactly true. When it encounters an edge of some sort, light diffracts (bends), just as ocean waves bend around the edge of a breakwater. This bending is more difficult to see with light than with water because the wavelength of light is so small. There are about fifty thousand wavelengths to an inch (twenty thousand to a centimeter). When light passes through a pinhole, the diffraction spreads it out to a spherical wavefront—picture it as half of an expanding balloon.

A short distance beyond the pinhole, Young put a second plate with two pinholes close together. He figured that they would pick up parts of the same wavefront from the point source. They would act as two point sources that would stay in phase (they would rise and fall together). At the far end of the table he put a screen to pick up the light. Sure enough, he got a curved pattern of dark and bright lines, called interference bands or fringes. It was similar to the two-stone pattern in water.

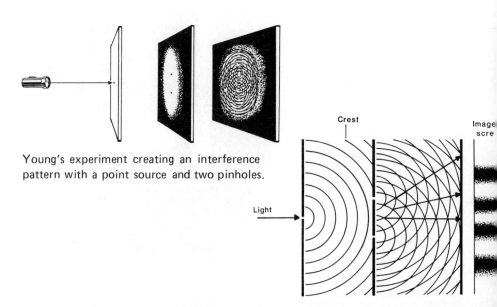

Young's experiment creating an interference pattern with a point source and two pinholes.

No one could explain that result in any other way: Light must consist of waves.

In a physics-laboratory demonstration of interference, the teacher generally uses narrow slits, rather than pinholes, in the plates. Slits also diffract light, and the wavefronts are shaped like half-cylinders rather than half-spheres. The interference fringes are straight, parallel bands of light and dark instead of the curved ones Young produced with pinholes.

Since the early 1800s people have accepted the wave nature of light. At first they imagined that all space must be filled with an invisible "ether"; that is, some material in which the waves form. However, no such material has ever been detected. Although light does pass through materials like air, water, and glass, it also travels through the vacuum of interstellar space. Evidently it doesn't need a material medium. Physicists today talk of light propagating by means of waves in an electromagnetic field, or of electromagnetic waves.

Color

Visible light is not all the same; it comes in a range of colors. Even the earliest people saw the magical beauty of rainbows in the sky.

Later, people examined light passing through glass prisms. They saw that white light (sunlight) goes in, and a rainbow of colors comes out. They concluded that the colors were somehow formed inside the prism.

Newton, however, did an experiment with two prisms that pointed in opposite directions. He showed that while the first prism formed a spectrum (spread the light into its colors), the second one recombined them into white light. Thus, the colors are not due to a property of the glass. The sunlight must contain them in the first place.

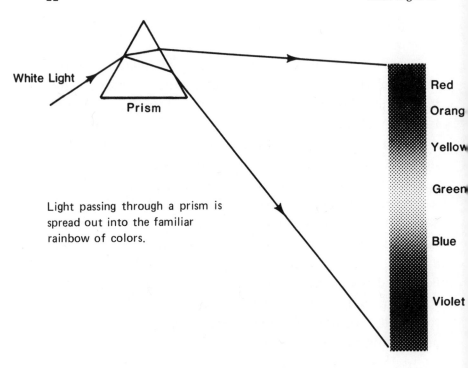

Light passing through a prism is spread out into the familiar rainbow of colors.

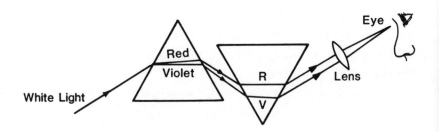

Newton used a second prism to recombine the colors, showing that they were contained in the original beam, not manufactured by a prism.

Once people had accepted the wave nature of light, they realized that each color has a different wavelength. The distance from crest to crest of light waves is very small for all visible colors, but the longest (dark red) have a wavelength about twice that of the shortest (deep violet).

Engineers generally talk about wave frequencies rather than wavelengths. The speed of the wave is equal to the frequency multiplied by the wavelength.

Imagine a wave traveling past you and count the number of crests passing per second. That is the frequency. Then measure the distance between crests to get the wavelength. The speed of the wave (the distance it has traveled in one second) is the number of crests that have passed in that second multiplied by the distance between crests.

All waves of the electromagnetic spectrum travel at the same speed in a vacuum; namely, the speed of light. Therefore, those of high frequency must have short wavelength.

The Electromagnetic Spectrum

That was not all. At about the time that Thomas Young was demonstrating interference, the astronomer Sir William Herschel (1738-1822) was looking at sunlight spread out by a prism. He put thermometers at various positions in the spectrum and found that the red end gave higher temperature readings than the violet end. To his great surprise, a thermometer placed beyond the edge of the red end, where he could see no light, showed a higher temperature than those in the light.

With some additional experiments he showed that whatever heated the thermometer could be reflected by mirrors and bent by prisms just like visible light. Herschel had discovered the infrared part of the spectrum.

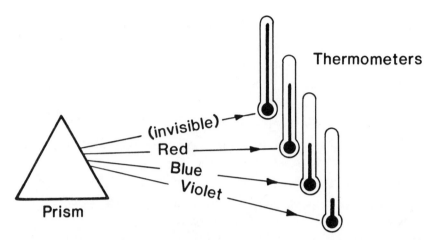

Hershel discovered the invisible infrared wavelengths when his thermometer placed beyond the end of the spectrum gave a higher reading than those in the light.

You know how it feels when you hold your hand close to a hot iron. Until Herschel's work, people hadn't realized that the heat they felt with their skin was similar to the light they could see with their eyes.

Gradually, scientists came to understand that there is a tremendous range of electromagnetic waves. They go from radio waves, measured in hundreds of meters, down through microwaves, infrared, visible light, ultraviolet and X-rays to gamma rays. They are all the same sort of thing—electromagnetic waves—differing only in wavelength. (If you prefer, you can say they differ in frequency.) What a spread of values it is! A typical radio wave has a wavelength about 10^{15} (1,000,000,000,000,000) times that of a typical gamma ray. Note that in order to get the illustration onto the book page, each wavelength division, going from top to bottom, has been made ten times the size of the preceding one. That is called a logarithmic scale. The frequencies are shown decreasing as the wavelengths increase. The unit, hertz, means vibrations per second.

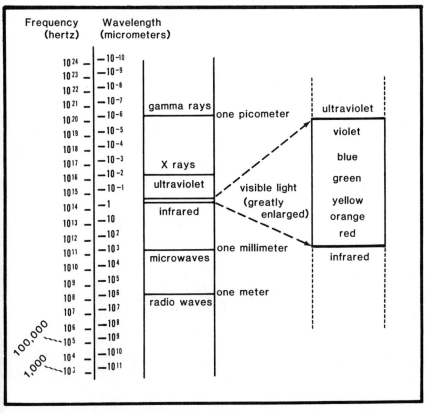

The spectrum of electromagnetic waves from very long radio waves, through visible to very short gamma rays.

You can see from the illustration that visible light is a very tiny part of the whole electromagnetic spectrum. Our eyes respond to just a small range of wavelengths.

Polarization

Another piece of evidence for light waves is polarization. Polaroid sunglasses are made of material in which the crystals have been carefully lined up in one direction. Put on two pairs of glasses, one over the other, and their crystals are lined up in the same way. You can see through the two pairs. Then take off one pair and turn it 90° (a quarter turn) in front of the other pair. Then try to look through both glasses with one eye. The crystals of the two pairs of glasses will be aligned crossways; little light will get through the double filter.

To understand what's happening, think of that vibrating rope again. Imagine two poles in the ground, one close to each side of the rope. An up-and-down wave would pass through easily, but if the wave were vibrating sideways, the rope would hit the poles and be stopped.

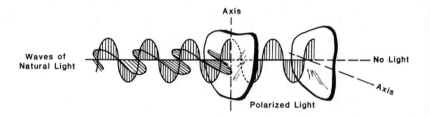

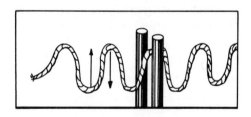

The aligned crystals in Polaroid permit only one direction of vibration— like a rope vibrating between posts. Add a second Polaroid lens turned 90°; together they stop all light.

Think of normal light as having both up-and-down and sideways waves. If the Polaroid permits only the up-and-down part to pass, then the light is said to be polarized in the vertical direction. Turn the second piece of Polaroid 90° and it stops the vertical waves as well.

Reflection from glass at just the right angle can also cut out one direction of vibration. So, although there was no Polaroid in the 1800s, scientists of those days were aware that light could be polarized. In fact they knew that light from parts of the sky is polarized and that this cannot be explained unless light consists of waves.

Some biologists now suggest that bees and birds may use this polarized light for navigation purposes. Airplane pilots are hoping for the development of a similar system for use near the earth's poles, where magnetic navigation is unsatisfactory.

The Photoelectric Effect

During the 1800s scientists were also learning a great deal about electricity and magnetism. Then in 1873 James Clerk Maxwell (1831-1879) published his comprehensive electromagnetic theory that unified light, electricity, and magnetism. It explained the earlier observations about light in terms of electromagnetic waves and predicted such properties of light as reflection, refraction, and interference. The wave nature of light now was well established, and people felt they understood these things. Newton must have been wrong. But was he?

In 1888 another phenomenon was discovered. Certain materials when illuminated by ultraviolet light, or even visible light, were found to emit some tiny charged particles called electrons. This was called photoelectric emission. The strange thing was that while the number of electrons kicked from the surface increased with the brightness of light, the speed of the electrons did not. The maximum electron speed depended on the wavelength of the light. No one could explain this in terms of electromagnetic waves.

In 1905 Albert Einstein (1879-1955) proposed the idea that light comes in photons (particles or chunks) whose energy is related to frequency—and therefore, to wavelength. The higher the frequency, the greater the energy. This means that photons of blue light are more energetic than photons of red light. (Energy is a measure of the amount of work the photons can do. Photons of equal energy knock electrons of equal energy—therefore of equal speed—off materials.)

Materials vary in how tightly they hold their electrons. For each material there is a certain minimum photon energy needed to kick electrons off the surface. Visible light may do the trick in some cases; more energetic ultraviolet photons may be needed for other materials. Making the light brighter does not change the photon energy, but it does mean that there are more photons, and therefore more ejected electrons. Einstein's photons explained the experimental results.

It's very strange—in some cases light has the properties of waves; other times it behaves like particles. Since physicists have to accept the experimental facts, they now talk of the dual nature of light—particles that come in waves. It appears that Newton and his opponents were all partly right.

The Speed of Light

Switch on a lamp and light seems to get to the farthest corner of the room instantly. Sound is obviously slower. You see the starter at a track meet fire a gun and hear the bang later. The boom of thunder often comes seconds later than the flash of lightning.

Light must travel very fast. Is it possible to measure the speed? That measurement was a real problem hundreds of years ago when there were none of the fast timing devices we have today. One possibility was to measure over very long distances.

In the 1600s, after telescopes had been invented, people became aware that Jupiter had moons that circled it in a regular manner, disappearing behind the planet and then reemerging. The Danish astronomer Olaf Roemer (1644-1710) timed several eclipses of Jupiter's moons and made a table predicting the times of successive eclipses. However, after some months his predictions became further and further removed from the actual times of the eclipses.

Roemer realized what was happening. It wasn't that a moon's motion around Jupiter was changing. Rather, as the earth traveled around in its orbit further from Jupiter, the light had further to go to reach his telescope, and the eclipse appeared to occur later.

He established that, though the speed of light was very high, it was not infinite. From the time lag that Roemer measured and the size of the earth's orbit as astronomers of those days knew it, the speed of light is calculated to be about 187,000 miles (300,900 kilometers) per second.

Years later Albert Michelson (1852-1931) obtained a more accurate figure. Today we have clocks that measure to billionths of a second and we can make measurements of the speed of light that are accurate to many decimal places. The accepted value now is 186,322.210 miles per second or 299,792.458 kilometers per second in a vacuum (slightly less in air).

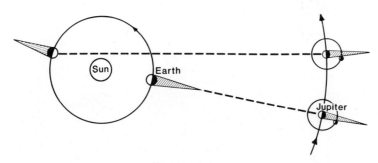

Roemer measured the speed of light by measuring the apparent difference in the time of eclipse of a moon of Jupiter. In one position (shown) the light traveled a longer distance by nearly the diameter of the earth's orbit around the sun.

All electromagnetic waves travel at the same speed in a vacuum. We have some experience with radio waves in this space age. When Ground Control called the Apollo 11 astronauts, the radio waves had to go some 239,000 miles (384,000 kilometers) to the moon and the answer had to travel the same distance back to earth. There was a tense 2.5-second pause before the famous words reached Houston, "Eagle has landed."

Traveling at 300,000 meters a second, it takes light from the sun about 8 minutes to reach the earth. We are 8 light-minutes from the sun. That's like saying that a city is "an hour away" at car speed. The planet Neptune is over 30 times as far from the sun as the earth is. It is more than 4 light-hours from the sun.

Distances from one star to another are so great that it is almost meaningless to give them in miles or kilometers. You get a better idea by talking about how many years it takes light to travel that far. The star nearest the sun is about 4 light-years away. With telescopes we can see galaxies of stars billions of light-years away. It is eerie to think that the light we now see from such stars left them before the earth even existed!

2
What Light Does

We now have some idea what light is: little photons traveling in waves at great speed. Visible light has only a small range of wavelengths (a small range of photon energies) but is part of a vast spectrum of electromagnetic waves.

In general, when these waves come in contact with material objects, they are partially reflected and partially absorbed, with some part passing completely through the material.

Imagine sunlight shining on a tree. You see the tree because some of the light is reflected from the leaves to your eye. In this case the green wavelengths are particularly well reflected, so the tree leaves appear green. The leaves absorb a large fraction of the other wavelengths and are warmed. Chlorophyll in the leaves—which gives them their green color—uses some of the absorbed energy to make food for the tree.

While very little light passes through a leaf, there are other materials quite transparent to visible light: glass, water, some plastics, and thin cloth. Some light gets through most materials if they are thin enough.

Reflection: Plane Mirrors

The word *reflection* probably makes you think first of looking at yourself in a wall mirror. That is a plane mirror (flat, all in one plane). Your image in a plane mirror is the same size you are, and is or appears to be just as far behind the surface as you are in front of it. It is also reversed, interchanging left and right. The reversal is confusing when you are fixing your hair or tying a tie. Your brain doesn't know which way to tell your hands to move.

Another example of a plane mirror is a very still lake reflecting trees, hills, and clouds above. However, if a breeze comes up and ruffles the water, the reflection disappears. Rough surfaces don't reflect like smooth metal or water.

Though these may appear to be different processes, it is satisfying to know that one basic law controls the reflection of light. It is, actually, the same as for the rebounding of a ball. The rule is that the angle of reflection is equal to the angle of incidence.

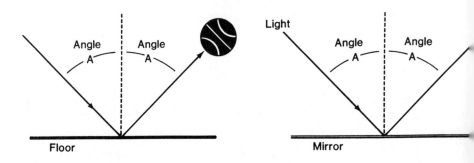

Both angles are measured with respect to an imaginary line perpendicular to the surface. The perpendicular line and the lines representing the direction of the incident and reflected light will all lie in the same plane.

What Light Does

If the angle of incidence is small so that the light is directed close to the perpendicular, the angle of reflection will be small. In fact, if the incident light shines along the perpendicular, the angle of incidence will be 0°, and the reflected light will bounce right back along the same line.

In the case of the lake, when the wind roughens the water, the surface becomes a whole series of little mirrors pointing in all directions. Sunlight obeys the law for each bit of surface. It reflects off at all angles. Your eye picks up all those confusing bits of light coming at different angles. There is brightness, but no clear picture. The reflection is diffuse.

Compared to the very short wavelengths of light, all but the smoothest surfaces are made up of many planes at different angles. That is true of this page of paper, your skin, the walls, the floor, and of most surfaces. They give diffuse reflections. You see most things because of the light they reflect diffusely to your eyes.

If you plan to build a system with lenses and mirrors, you will want to be familiar with the details in the next three sections. However, if you want just an overview of optics, you may prefer to skim through to the section on spectroscopy. Refer to the glossary for definitions of the terms used.

Reflection: Curved Mirrors

Have you ever seen your reflection in the side of a shiny new car? You can look short and fat, elongated, or weirdly distorted as you move around. The car is not a flat, plane mirror but has various curved surfaces. Light still follows the law that the angle of reflection equals the angle of incidence, but the net effect is to make the image a different size and shape from the object.

People who design mirror systems need to know how light will move through a system and what kind of image will be formed. They get a quick idea by using ray diagrams. This is not to imply that light travels in rays. These imaginary rays represent the direction the light waves will travel as they go to and from the mirror.

Sometimes mirrors turn the image upside down. By making the object an arrow and drawing rays from the arrowhead, it is easy to see if the image is inverted. For the next few examples let's assume we are using spherical mirrors (shapes cut from a large sphere or ball). That's the shape most often used in optical instruments.

A concave mirror is like the inside of a spoon (except that the one illustrated is part of a sphere). A ray that goes through the center of the sphere will be perpendicular to the mirror. Its angle of incidence will be 0°; therefore, it will reflect back along the same line.

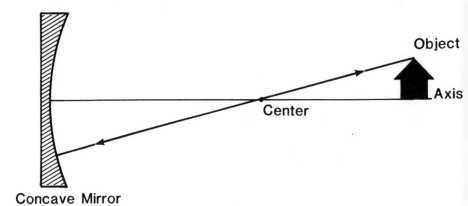

Concave Mirror

Note the line labelled axis. You could rotate the mirror around the axis (like a wheel on its axle) without changing anything. Rays that are parallel to the axis, as though coming from a great distance, will reflect to points close to the focus of the mirror. You could find the focus experimentally by seeing where light from the sun collects. It turns out to be halfway between the mirror and the center of the sphere.

We could spend a lot of time drawing rays to various parts of the mirror from all parts of the object, always making sure that the reflected and incident angles are the same, but would learn no more than by drawing just the ray through the center and the parallel ray from the tip of the arrow. Where they cross is where the image of the tip of the arrowhead will be.

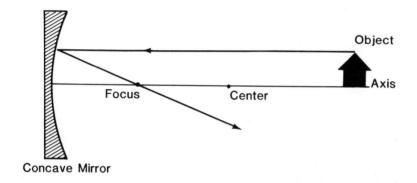

Concave Mirror

In the figure below the object is well away from the mirror, outside the center of curvature. That gives an image which is smaller than the object, is inverted, and lies between the center and the focus. It is a real image because the light focuses. You could put a piece of film at that place and get a picture of the object. Even if it looked funny and distorted, it would still be called a real image.

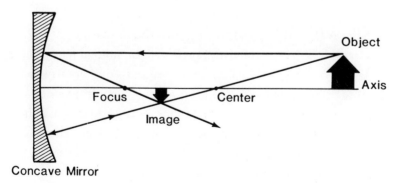

Concave Mirror

Now turn this spherical spoon over and reflect light from the convex outer surface. A ray aimed toward the center of the sphere (which is behind the mirror in this case), will bounce back along that same direction. The dotted line in the drawing continues that ray behind the mirror to the center.

A ray along the direction of the axis will bounce off the mirror as though it had come from the focus. A dotted line extends that ray behind the mirror also.

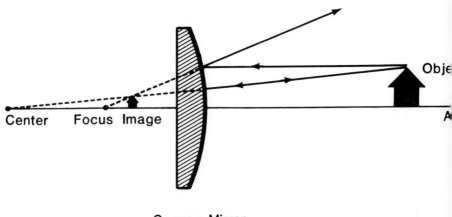

Convex Mirror

Notice that the rays reflected off the mirror are diverging. They will never come together; there will be no real focus. Your eye lens, however, will gather them up and your brain will tell you that they came from the point behind the mirror where the dotted lines cross. You'll see the image even though it is behind the surface. It is a virtual image with no real focus. There is no place you could put a piece of film to record a picture of the image without using a focusing lens.

What Light Does

The image is upright, small, and behind the mirror. You may have seen that same effect in one of those reflecting lawn balls.

Spherical Aberration

Spherical mirrors can cause problems of blurring. Light reflected from different parts of the mirror passes through a region all around the focal point rather than through the focal point itself.

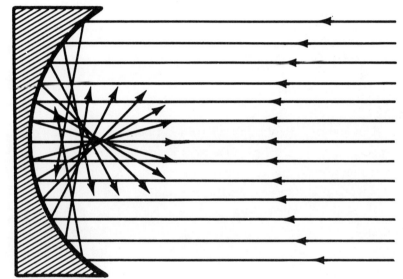

Spherical Mirror

This effect, called spherical aberration, is a result of the round shape and of the law of reflection. It would be a serious drawback in a telescope mirror, causing the blurring of star images. The blurring is minimized if the mirror is only a small fraction of the surface of a large sphere. However, there is another shape, called paraboloid, which is preferred for telescope mirrors (and for satellite dish antennas). It does focus parallel rays to a single point.

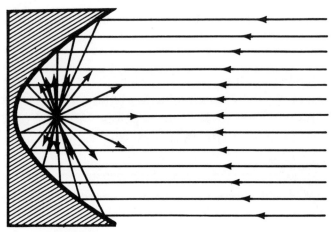
Parabolic Mirror

The law of reflection is reversible; it holds whether the light is going in or out. Therefore, rays from the focal point of a parabolic mirror are reflected out parallel to the axis. That property is used in a searchlight. A small bright source placed at the focal point will be reflected as a parallel beam. In actual practice, the entire beam is not exactly parallel. The source is bigger than a point, and so most of it is not quite at the focus. There is also some spreading of the beam due to the diffraction (bending) of light mentioned in Chapter 1. Diffraction, you recall, is something that happens to waves when they hit an obstruction. We're now going to talk about another type of light bending called refraction.

Refraction

Though today's largest telescopes use mirrors to focus starlight, the earliest telescopes used glass lenses. Many small telescopes still do. Lenses work because of the way materials affect light passing through them. A clear material like glass doesn't absorb much of the visible light but does bend it.

A light wave striking a glass plate at an angle is bent inward toward the perpendicular. When it gets to the back surface and emerges into the air, it bends outward again. The photons have interacted with the atoms of the glass. Though they emerge unchanged, the interaction slows them. It is that slowing that causes the bending.

By analogy, imagine driving a car into a strip of deep sand across a road. If you head straight into the boundary between pavement and sand, the wheels on both sides slow together and the car goes slowly straight forward.

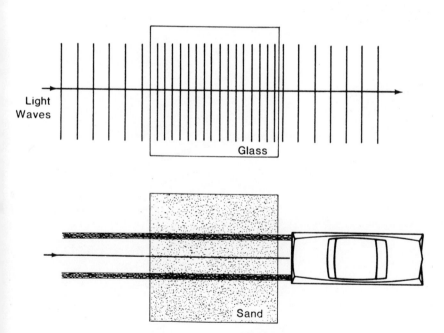

If you approach at an angle such that both right wheels reach the sand first, the right wheels slow first and the car swings right. Once both right and left wheels are in the sand, the car will slowly continue in the new direction. At the end of the sand strip, the right wheels emerge first and start going faster. The car veers left.

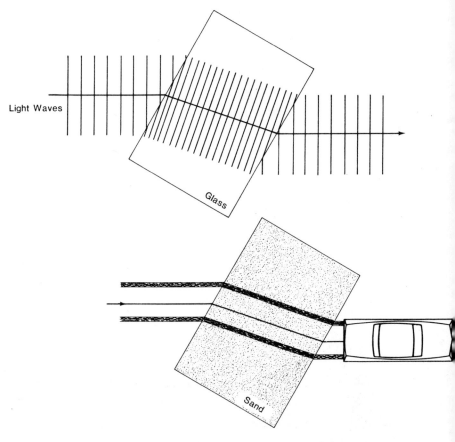

Light is refracted when it goes from one medium to another. The boundary may be between air and glass, air and water, or any two media—even hot air and cool air. The angle of refraction depends on the two materials as well as the angle of incidence of the wavefront.

What Light Does 41

You can easily show yourself an example of refraction at an air-water boundary. Place a coin in the bottom of a bowl. Raise the bowl to a level in front of your face such that the coin is just hidden behind the rim. Then pour a glassful of water into the bowl and watch the coin come into view.

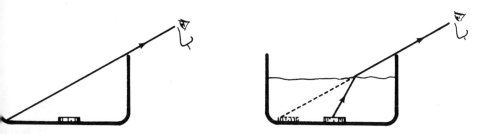

That is a case of refraction at a plane boundary between materials. Refraction also occurs at curved surfaces. A magnifying glass, with its two curved surfaces, is thicker in the middle than at the edges. It is a convex lens. Parallel rays from the distant sun can be brought together at a point on the opposite side of the lens. No doubt you've used a magnifying glass to char holes in paper placed at the focal point.

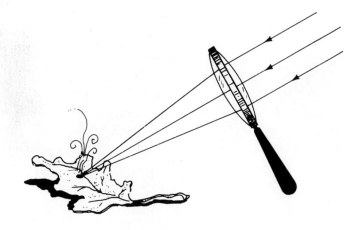

There are simple ray diagrams for lenses, too. A ray through the center of the lens will continue without bending. A ray parallel to the axis will pass through the focal point. (As with mirrors, the axis is a line around which you can rotate the lens without changing anything.)

Any object further away than the focal length of a convex lens will form a real inverted image on the opposite side of the lens.

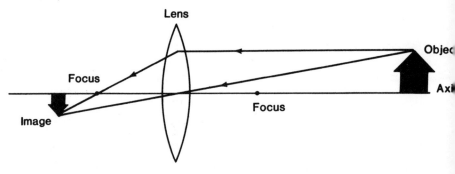

However, if you hold the magnifying glass close to the printed page inside the focus, light reflected off a letter makes such a sharp angle with the glass that, after passing through the lens, the rays are still diverging. Your brain extends the rays back, like the dotted lines in the picture, and says they came from a large letter behind the page. You see a magnified virtual image.

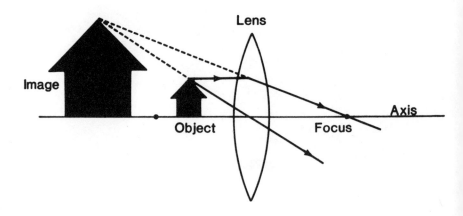

What Light Does

Lenses are made in a variety of shapes; one side may be flat while the other is convex or concave. Both sides may be curved. If the center is thicker than the edges, the lens will converge light incident on it.

Concave lenses, which are thinner in the middle, tend to spread light. They can be very useful as eye glasses. In nearsighted people, the eyeball is too long and the muscles can't adjust the eye lens properly. The image of a distant object forms in front of the retina. A concave, diverging lens can move the image back to the retinal surface.

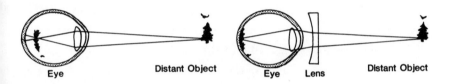

The opposite is true for farsighted people. Their eyeballs are too short, and the image falls behind the retina. A convex, converging lens brings it forward.

Things get more complicated if the eyeball is not round. The eye lens may not focus vertical objects in the same way that it does horizontal ones. That is astigmatism, and it may require a cylindrical lens rather than a spherical one to correct the vision. An eye doctor can test to find out what shape lens is necessary.

We have treated lenses as though they slow and bend all wavelengths the same amount. That is not quite true. The violet wavelengths are slowed more than the red and are bent further. Different colors focus at different distances. This chromatic aberration is small with thin lenses and can be reduced further by the proper choice of lens materials.

When it is desirable to spread out, or disperse, light, people can use prisms. They choose a wider prism angle to increase the dispersion. It is also possible to choose a glass that spreads red light and blue light apart to a greater degree than does ordinary glass.

Spectroscopy

In Chapter 1 we saw how a prism disperses white light into its component colors. Scientists in the laboratory have designed instruments, called spectroscopes, in which light from an incandescent material passes through a narrow slit and then through a prism (or another device called a grating). The light, spread out according to color, falls on a calibrated screen or film. Focused by a lens, each wavelength of light present in the beam forms a bright, narrow-line image of the slit on the screen. The experimenter can read the wavelength on the scale.

A glowing metal, as in an ordinary light bulb, emits a continuous spectrum, one color shading into another. A low-pressure gas lamp, on the other hand, emits just a few sharp lines.

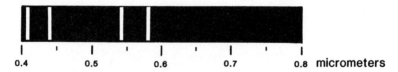

A bright-line spectrum of the element mercury.

Each chemical element produces a certain characteristic set of wavelengths. Long years of experience have enabled chemists to make a dictionary of those lines. They can use each set to identify the element that made it; for example, hydrogen, oxygen, or iron. Scientists have also learned to recognize whether the materials are at extreme temperatures and whether they are in strong electric or magnetic fields.

What Light Does

Furthermore, if light from a continuous source shines through the vapor of an element, that element will absorb light at its characteristic wavelengths, leaving dark lines in the continuous spectrum.

All this is important to astronomy. A careful study of sunlight with a spectroscope shows a vast number of narrow, dark lines superimposed on the familiar rainbow. Light from the very hot interior of the sun is absorbed by atoms in the somewhat cooler outer layers. By studying the wavelengths of these dark Fraunhofer lines, scientists have been able to learn what elements are present in the sun. In fact, the element helium was discovered from its lines in the sun's spectrum before it was known on earth. It was named after Helios, the ancient Greek sun god.

Astronomers look at stars so distant they aren't visible without a telescope. Amazingly, they can tell what those stars are made of, how hot they are, and how fast they are moving away from, or toward, us—all from a study of their spectral lines.

Having considered basic optics, let's now go on to the newer developments.

3
Coherent Light: The Laser

Almost everyone has heard of lasers. Some know that the name is made up of the initials of the words *L*ight *A*mplification by *S*timulated *E*mission of *R*adiation, but few have seen a laser. Though they are used in laboratories, hospitals, in the space program, and in industry, lasers are not common household items. To see several types of lasers in one place, you must go to a science museum or a large industrial laboratory.

A laser is a light source. So are the sun, an incandescent bulb, a fluorescent tube, and a candle, to name a few. The difference is that while the others emit light in all directions with a great range of wavelengths, the laser puts out a narrow beam of light of a single color at a time. Furthermore, all its waves rise and fall together. It is called a coherent source. Coherent light, it turns out, can do things different from what ordinary light can do.

Energizing the Atom
All visible light comes from electrons jumping around in atoms and molecules. You can picture an atom as a miniature solar system. A dense nucleus takes the place of the sun, with a series

of electrons, like the planets, orbiting around it. That isn't exactly accurate, but it gives a convenient mental picture.

Pictorial representation of an atom with electrons in orbit around a dense nucleus.

Physical laws that apply to large objects like suns and planets don't work for such tiny ones as electrons and nuclei. A whole new theory, the quantum theory, was developed to deal with submicroscopic particles.

According to quantum theory, electrons in an atom can't have just any energy. There are certain allowed energy levels. It's like climbing a ladder. You have to raise your foot all the way to the next rung; you can't step part way. The lowest energy level of an atom is called its ground state. When an electron has been lifted to a higher level, it is in an excited state.

To raise an electron from one allowed level to another it is necessary to put in an amount of energy almost exactly equal to the difference in the two levels. There are several ways of getting energy into atoms. In a light bulb it's electrical. Fireflies use chemical methods. Atoms in the sun and stars get their energy from nuclear reactions.

Another method is to aim light at the atoms. Newton said that light comes in small particles, and the photoelectric effect shows that light sometimes behaves that way, even though it behaves like waves under other circumstances. When a particle (photon) of light strikes an atom, it may disappear. An electron gains all the photon's energy and rises to a higher level. It cannot take just part of the photon energy.

Coherent Light: The Laser 49

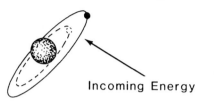

Photon or other incoming energy of the right amount kicks an electron from dotted line orbit to higher energy orbit.

If no permitted level exists at the higher energy, the light will not be absorbed, but will pass through the material with all its photons intact. Clear glass is transparent to light because photons of visible light don't have enough energy to raise electrons to the next allowed level.

Colored glasses, on the other hand, contain small amounts of metallic elements added to the silica, soda, and lime of clear glass. Those extra atoms provide different energy levels which absorb some of the visible photons. If they absorb the blue end of the spectrum, only the longer wavelengths pass through and the glass looks red.

Even clear glass will stop ultraviolet photons. They are sufficiently more energetic than photons of visible light to raise electrons to allowed levels. Ultraviolet light doesn't affect your skin through a closed car window, but if you open the window and take a long, sunny drive with your elbow on the ledge, you'll have a tanned arm. Fortunately, the earth's atmosphere absorbs the high energy end of the ultraviolet range and so shields us from the really deadly photons.

Very energetic photons can kick an electron completely out of an atom. That is called ionizing the atom. For each material there is a certain minimum energy required to cause ionization. All photons with more energy than that minimum are absorbed by the atom and boot electrons out of the material.

When electrons have been raised to excited levels, they will spontaneously drop back to the lowest level—the ground state. As they do, the atoms emit the energy difference as an electromagnetic chunk. For a low energy, this chunk will be an invisible infrared photon; for higher energies, visible or ultraviolet photons. In some cases the electron may cascade down through several levels, emitting a photon at each step.

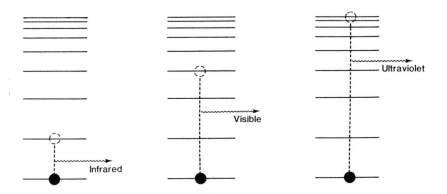

An electron falling to the ground state of an atom from a low energy level emits an infrared photon; from a medium energy level, a visible photon; and from a high energy level, an ultraviolet photon.

Coherent Light

In the usual light source, billions of atoms have been excited to a variety of higher levels. As they drop back, they emit photons at random times in random directions. The light that comes out is called incoherent. The light waves have many different wavelengths and go off every which way, out of step, like students coming out of the building when school is over.

Coherent light, on the other hand, is like a champion school marching band: in line, all taking the same size steps, all keeping time. They are in phase. Laser light is coherent.

Coherent Light: The Laser

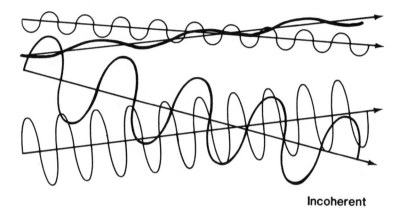

Incoherent

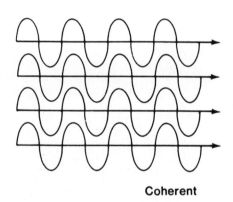

Coherent

From an ordinary source, light waves are emitted with random direction, wavelengths, and phases. From a laser, the waves are coherent.

Scientists were already familiar with coherent radio waves, which are routinely used for broadcasting. But those are long waves, measured in meters. A light wave has a wavelength of only about 20 millionths of an inch (one-half millionth of a meter). How can you control atoms too small to see, and make them all emit light of the same color, in the same direction and all in phase?

Let's use an analogy that seems like a weird dream. You are playing with a beach ball on some old, uneven steps where birds have nested. You are throwing the ball up the steps. The more energy you put into the ball, the higher the step it will reach. It immediately bounces back down, but you've disturbed a bird on that ledge which flies off as the ball falls.

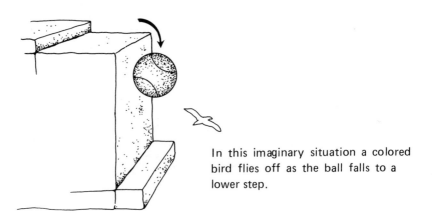

In this imaginary situation a colored bird flies off as the ball falls to a lower step.

Electrons excited to higher energy levels in an atom usually drop back immediately to the ground state. As they fall, a photon is emitted carrying the energy difference. If the level is high, the photon will be at the blue end of the spectrum. From low levels, the photons will be red. Imagine blue birds on the upper steps, red ones on the lower steps, and the other spectral colors in between.

It is important to understanding laser action to note that some steps are wider than others. The ball will stay on those a lot longer before falling.

Stimulated Emission

Another fact basic to laser operation is something that Albert Einstein calculated when studying the emission and absorption

Coherent Light: The Laser

of light in atoms. Physicists had known about photons being absorbed to raise electrons to higher energy levels. They had known about photons being emitted as the electrons fall back. Einstein found that there is another possibility. If an electron is resting on a level at a given energy and a photon of that same energy comes along, the photon will not be absorbed but will nudge the electron so that it falls and emits another photon. Then there will be two photons instead of one. Einstein called that stimulated emission.

He calculated that the second photon will have the same energy, or color, as the first. Furthermore, it will come off in the same direction and in phase with the first photon.

In our analogy, the beach ball is precariously balanced on a step. A bird flies by and the breeze of its wings makes the ball tumble. That scares up another bird. The second bird is the same color as the first, flies in the same direction and beats its wings in time with the first. In other words, the birds are coherent. Coherent radiation is characteristic of laser operation.

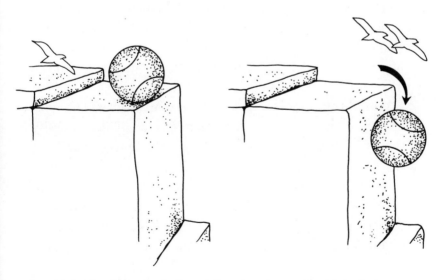

One bird stimulates the ball to fall and emit another bird of the same color, phase, and direction.

Amplification

Einstein's calculation covers stimulated emission of radiation (the S-E-R of L-A-S-E-R). There must also be light amplification.

Normally almost all the atoms of a material have their electrons in the lowest energy states. Most photons will hit those and be absorbed. They won't stimulate other photons. To get amplification, it is necessary to have the majority of atoms in the same excited level (many beach balls poised on the same step). Stimulated emission can then lead to a group of coherent photons.

A special kind of excited level is needed. It must act like a wider step, so that the electrons will pause there before falling back. Though they may pause for less than one thousandth of a second, that is still millions of times as long as an electron spends in an ordinary excited level. Physicists call such a long-life level a metastable state.

Not all materials have such a wide level—at least not at an energy to provide a visible-light photon. In 1960 (applying theory developed by Charles Townes and Arthur Schawlow), Theodore Maiman at Hughes Research Laboratories made the first operating laser. He used a ruby crystal. Ruby is basically aluminum oxide with a small amount of the element chromium giving it its red color. The chromium also gives it a metastable energy level.

To excite atoms to the necessary state, Maiman used photons of a range of energies from a flashlamp. A burst of high intensity light kicked electrons up to a series of high energy states. In falling back, a large fraction of them landed on the metastable level. During the time they rested there, there were actually more atoms in that excited state than in the ground state. Since that is just the opposite of the normal situation, it is called population inversion. Maiman got population inversion in his ruby crystal.

Coherent Light: The Laser

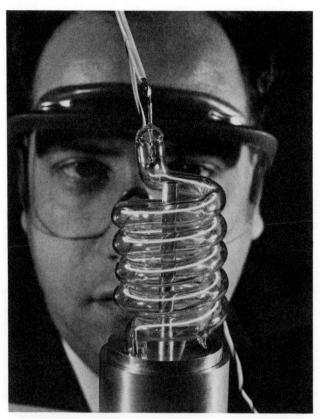

Dr. Theodore Maiman of Hughes Research Laboratories studies the laser's main parts, a spiral flashlamp surrounding a rod of synthetic ruby crystal.

After a very short while, by human standards, one of the electrons fell by chance. Its photon went past another excited atom and stimulated it to emit. Then there were two coherent photons. They stimulated two more atoms for a total of four photons, and so on—a chain reaction.

By the time they reached the end of the crystal, their number was large—though still far from the total number of available photons. Maiman, however, had polished both ends of his ruby so they were accurately flat and parallel. He coated them with aluminum to make good reflecting surfaces. When the photons reached one end, they bounced off and started back, stimulating more photons to join them in the same direction.

Bouncing back and forth, they raced up and down the ruby, rapidly multiplying photons, amplifying the beam. Because light travels so fast, it can make a round trip through a 4-inch (10-centimeter) long crystal in one billionth of a second. The laser light isn't useful if it stays inside the ruby, so Maiman made the reflective coating on one end thin enough to let about 10 percent of the light through on each pass.

Of course, some spontaneously emitted photons went off not down the length of the crystal but out through the side walls. They were lost to the beam. Only those surging back and forth between the mirrors were included in the burst of light coming out the end in a narrow pencil-like beam. The longer the crystal, the more narrow the beam.

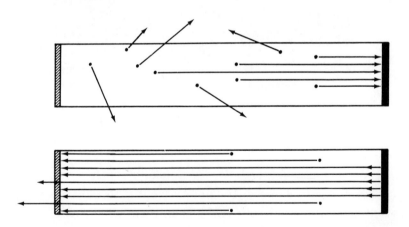

Photons emitted through the side walls of the laser are lost to the beam. Those traveling the length of the laser stimulate other photons to join them as they race back and forth between mirrors. On each pass some escape through the partial mirror on the left.

Coherent Light: The Laser 57

The laser beam spreads very little, even at great distances. To measure the earth-moon distance, short pulses from an improved ruby laser have been sent to the moon and bounced back by reflectors installed there by the Apollo astronauts. That's a round trip of about a half million miles (800,000 km)! The fact that scientists can measure the return signal shows how little the light spreads.

Though Maiman's ruby laser was the first laser to operate, other scientists who had been attempting to make gas lasers succeeded soon afterward. In laboratories all over the world, people began trying different materials. They achieved lasing with a great number of solids, liquids, and gases. In each case they had to excite atoms or molecules to achieve population inversion. In general, they used mirrors to form the resonant cavities to reflect the light back and forth. The techniques varied with the materials.

In the early 1960s, some engineers liked to call the laser "a solution in search of a problem." You no longer hear that. Let's look at some of the many jobs lasers are doing. There are problems for which they are the best or the only solution known.

4
Laser Light in Use

As we saw, Theodore Maiman used a burst of intense light to excite the chromium atoms to high energy states in his ruby laser. That's called optical pumping. Optical absorption is not, however, an efficient way of exciting the widely spaced atoms of a gas.

Gas Lasers

Most gas lasers are pumped electrically by applying a high voltage across a tube of gas. A current of energetic electrons then flows through the gas. The electrons bang into atoms and transfer energy to them.

The most common of the gas types is the helium-neon laser. Electrons excite helium atoms to metastable levels which happen to have almost exactly the same energy as two levels of neon. When the atoms collide, the helium excites the neon atoms to those levels. It is the neon which then provides the laser light.

While the ruby laser is operated in bursts with cooling time between pulses, it is possible to run the helium-neon laser continuously if a steady voltage is applied. That is called a continuous wave, or c.w., laser.

The helium-neon laser is the least expensive of the gas lasers. Its red light is familiar to students, laboratory technicians, surveyors, people aligning tunnels or drainage systems, and to doctors doing laser acupuncture. It is also used at the supermarket checkstand where laser light reflects off the universal product code (that little pattern of dark lines) on packages. You hear a *beep* when the computer recognizes the code and causes the item's name and price to be printed out on the tape.

Neon is not the only rare gas useful in lasers. The green light from argon is well absorbed by blood and is often used to repair detached retinas in eyes. Doctors have also had some success in clearing up the tiny eye hemorrhages that can cause blindness in diabetics.

Both argon and krypton have not one but several emission lines in the visible region of the spectrum. Systems using them can be designed to use a single color or several colors simultaneously.

Dr. A. Stevens Halsted of Hughes Aircraft operating an argon laser. Its several wavelengths in the blue-green region of the visible spectrum are spread out by a grating.

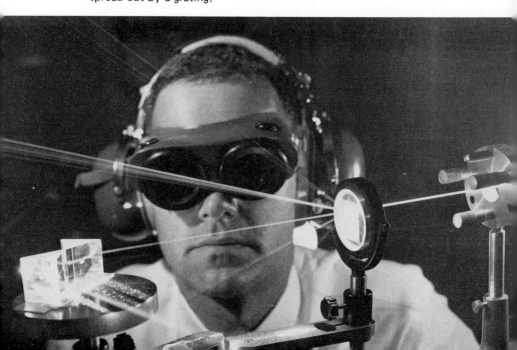

Laser Light in Use

Perhaps you have been entertained by a light show. You might have seen laser beams of several colors sent through smoke or fog to make them visible. The colored beams form patterns by reflection from mirrors that rotate or vibrate in time to music. These shows have been popular for some years and are still performed in some museums and theaters.

As ocean water is quite transparent to blue-green wavelengths, the argon laser is also useful for communication signals underwater and between submarines and satellites.

You may know that in a digital computer all numbers and instructions are encoded into strings of 0's and 1's. Music and speech can also be put into digital signals. One of the advantages of the narrow coherent beam of a laser is that a lens can focus it to a microscopic spot. Light focused from an argon laser can be used to form millions of tiny depressions in a spiral pattern on a disc, like the groove of a phonograph record. These represent music in digital form.

Read by a low-power laser, the signal is converted by the player into high-quality sound. As long as the readback device can distinguish between a hollow and no hollow, it will be accurate. These recordings are much less subject to damage in handling than ordinary records are, and there is no needle wear.

Because the dots are so tiny, it is possible to record information, not only for music, but for the far more complicated TV signals. Laser video discs and players compete in the marketplace with video cassette recorders that use magnetic recording. Laser discs may also become important for storing computer data.

Carbon Dioxide Lasers

Most gas lasers are quite inefficient. Only about 1 percent of the electrical energy put into them comes out as light. In the carbon dioxide laser, however, this number is more like 25 percent, and it can deliver high power. It is easy to see why this laser is widely used, even though it emits at 10.6 micrometers (μm), well into the infrared region.

Though also a gas laser, the carbon dioxide one is different in that the energy levels are not of excited electrons but of oxygen atoms vibrating in the carbon dioxide molecule.

The carbon dioxide laser is the one most used for laser surgery. The long wavelength of its beam is readily absorbed in water, which makes up the major part of body tissue. With a lens to focus the laser's parallel beam to a very small spot, it can make neat cuts by vaporizing a fine line of tissue. It also cauterizes (heat seals) the blood vessels as it goes, so there is little bleeding.

A medical laser is large and heavy. Its infrared beam is directed by a series of mirrors on a flexible arm, much like the arm of a dentist's drill, so the doctor can move it around. Since the surgeon can't see the 10.6-μm wavelength beam, light from a helium-neon laser is sent along the same mirrored path to make a visible spot.

In industry the carbon dioxide laser is used for heat-treating metals to harden them. It drills holes in plastics and in hard materials like ceramics and diamonds. Whereas an ordinary drill would deform the rubber, this laser makes perfect holes in baby-bottle nipples. The clothing industry uses it for automated cutting of stacks of cloth.

High-power carbon dioxide lasers are used for welding thick metals, low-power ones for micro-soldering tiny electronic circuits. When the coherent light is focused down to a width of a few micrometers, enormous power is delivered to a very small spot without heating and warping the surrounding material.

Chemical Lasers

Although there is some military use of very high power carbon dioxide lasers, military researchers have looked more to lasers using chemical reactions, rather than electric currents, for their

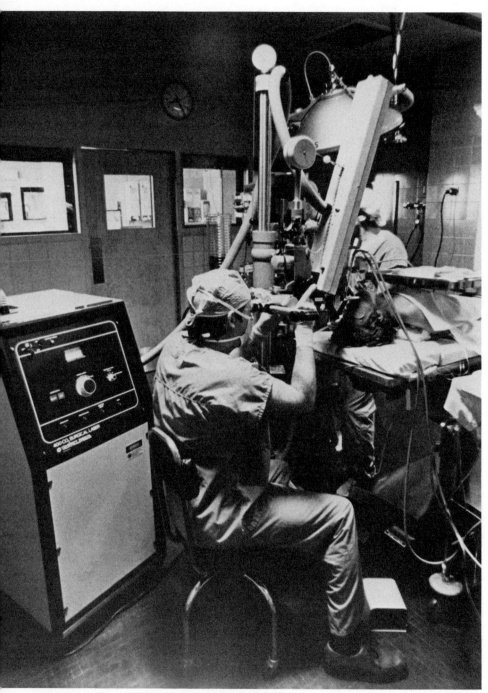
A surgeon using a medical carbon dioxide laser.

A laser processing center: carbon dioxide laser cutting head over a fast-moving cutting table.

pumping energy. Some laser antimissile weapons under development are powered by combining hydrogen and fluorine gases. They roar like a jet engine and use mirrors a couple of feet across. Now there is doubt that even these powerful lasers can do the job.

There is disagreement on whether lasers can make good weapons. At first glance, they seem to be the death rays of science fiction, but closer inspection shows that this isn't really true. Even the greatest enthusiasts admit that lasers are less efficient people-killers than bullets.

There are some obvious advantages to lasers: speed-of-light travel, rapid beam shift from target to target by the use of mirrors, accuracy of sighting, and strong focusing. The Army, Navy, and Air Force have each been putting millions of dollars into laser research.

The disadvantages are less obvious. High power carbon dioxide lasers are huge and need massive auxiliary equipment. Hydrogen fluoride lasers are powerful, but the gases are dangerous to handle. In any case, in the atmosphere light beams can be absorbed by clouds, fog, and water vapor, by atmospheric smoke, and conceivably by enemy smoke screens. Mirror surfaces put up by the enemy could deflect laser beams with little damage to the target. Aiming is difficult because high power lasers heat the air they pass through, changing the index of refraction of the air in such a way as to spread and deflect the beam. Dust or dirt on the mirrors of the laser system could cause them to absorb so much energy that the laser would self-destruct.

Outside the atmosphere, the problems are not as great. Military planners talk of putting into orbit laser battle stations that could shoot down intercontinental missiles before they can reach their targets. Even in space, there would be problems, however. Unlike an explosive weapon, a laser has no effect if it just gets close. It must hit exactly to effect a "kill." Furthermore, there is no quick way to know whether a missile has been hit.

Though the armed forces aren't using laser weapons now, they do use lasers for range finders and for computerized fire control of traditional explosive weapons. They also illuminate targets with an infrared laser beam, and have "smart bombs" which home in on the spot. The military uses laser radar and is working on laser communication between satellites and submarines.

Liquid Lasers

Most lasers emit a single wavelength, or perhaps a few isolated wavelengths. Some scientists, however, need a continuous spectrum of intense coherent light. For instance, a photochemist may want to study the effects of small changes in wavelength on chemical reactions. Liquid lasers—organic dyes in liquid solution—can provide such a continuous spectrum.

Organic dye molecules are complex, containing dozens of atoms. They have a large number of closely spaced energy levels and can be "tuned" to a particular wavelength. Instead of having a mirror at one end, a tunable laser has a prism which is rotated to reflect back the desired wavelength and stimulate emission at that color.

One dye will cover only a small portion of the spectrum, but by using a selection of the many available dyes, the researcher can get wavelengths covering the whole range of interest.

Solid Lasers

Ruby lasers are still in use, particularly where short pulses are needed. A trick for getting very short pulses is to prevent any of the light from getting out during the time that increasing numbers of photons are reflecting back and forth. In effect, totally reflecting mirrors are placed at each end of the ruby. Then one end is opened, and all the light bursts forth in, perhaps, a millionth of a millionth of a second.

Short pulses are useful for making holes, as in the photograph. They are also good for determining distance by accurately measuring the length of time a pulse needs to travel the distance and reflect back.

Ruby is used only for pulses because it gets hot and would change shape if used continuously. Without cooling time, the end mirrors would get out of adjustment. However, there is another crystalline material—called YAG because it consists of yttrium, aluminum, and garnet—that conducts heat away so well it can be used for a continuous beam. The metastable level in a YAG laser comes from a small amount of the element neodymium. Surgeons have found this laser particularly useful for treating bleeding ulcers.

Laser Light in Use

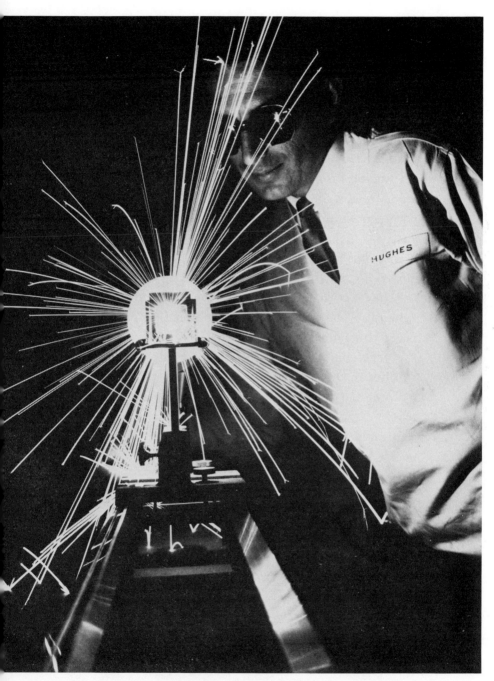

A ruby laser piercing a hole through a sheet of extremely hard tantalum.

All crystalline materials must have their atoms precisely placed if they are to have good optical properties. It is difficult to grow large crystals with the necessary perfection. Ruby and YAG lasers are fairly small. But researchers were happy to find that they could form large lasers with different kinds of glass. One important glass laser contains neodymium, the same element used in YAG. Some glass lasers are truly gigantic, for example, those used in fusion experiments.

Nuclear fusion is the process in which the nuclei of hydrogen atoms fuse, or join together, to form helium. A great deal of energy is emitted. This is how the sun and stars generate their heat and light. On earth, however, without the tremendous gravitational forces of a star, it is very difficult to keep hydrogen nuclei close together long enough for them to fuse.

One experimental method is to concentrate a huge burst of light from many arms of a high power laser on a pinhead-size pellet containing hydrogen isotopes. The light exerts forces on the pellet, and it must come in uniformly from all sides or it will just bat the pellet around. If the laser light is sufficiently uniform and intense, it will squash the gas down to about one ten-thousandth of its original volume and should produce fusion. The problems are enormous, but if the project is successful, nuclear fusion may one day supply unlimited energy.

Semiconductor Lasers

Semiconductors are a class of materials which conduct electricity somewhat better than insulators such as plastics, but not nearly as well as metals such as copper. Unique electronic properties make semiconductors useful for the transistors in radios and calculators. Some semiconductors also emit light and under just the right conditions can be made to emit coherent light; that is, they become lasers. Really tiny lasers, some as small as a grain of salt, are made of the semiconductor gallium arsenide.

The Nova system completed in 1985 consists of 10 laser arms which fill a large building. Their total power (over 100 million, million watts) is concentrated on a tiny pellet.

It is beyond the scope of this book to describe the energy structure of semiconductor crystals. We can say, however, that like other lasers, semiconductor lasers require population inversion and stimulated emission. Pumping is achieved by an electric field and the efficiency is very high—perhaps 75 percent. The total power output of such a tiny device is, of course, small.

Miniature semiconductor lasers are very important in the field of communications, now that communications is turning more and more to optics. In the next chapter we will discuss the optical-fiber communication systems that employ semiconductor lasers.

A tiny semiconductor laser in the eye of a needle.

Safety

When focused by the eye lens, the concentrated, coherent light from a laser can do permanent damage to the retina. It is important to be especially careful when working with invisible infrared or ultraviolet wavelengths.

If you have a chance to use a laser, be sure to wear protective glasses. Never look directly into a laser beam, even with protective lenses on, and also be careful of laser light reflected from surfaces around you.

5
Light in Fibers

The idea of optical communication is not new. Indians used smoke signals, navy men on ships have signal flags, and Boy Scouts use mirrors to reflect sunshine. Before inventing the telephone, Alexander Graham Bell tried to make a "photophone" that would send light signals through the air. Clouds, smoke, and atmospheric disturbances made that too unreliable. Bell knew it wasn't possible to build tunnels for long-distance light transmission and, until recently, there was no material that didn't absorb light within a few feet. Bell was forced to settle for electrical signals in wires—our familiar telephones.

Electrical communication has worked very well, but the demand for telephone lines continues to increase. Not only do businesses and individuals make more calls, but there is an increasing need for "talk" between computers. Since optical frequencies are billions of times greater than radio frequencies, optical signals can carry billions of times as much information. Hard-pressed telephone companies have longed to use optical methods, but what could they use for a light source and what could serve as a medium for transmission?

The invention of the laser, particularly the tiny semiconductor laser, provided a suitable light source. Now there is the astonishing development of glasses so transparent to short infrared wavelengths, and so pure and flawless, that they can transmit light with little loss for a hundred miles. British and American manufacturers have found methods for coating rods of such glass with material with a low index of refraction and then drawing them out into hairlike fibers. Fibers are effective because of the phenomenon called total internal reflection.

Loops of hair-thin glass fiber, illuminated by laser light, represent a fiber optic system.

Light in Fibers

Total Internal Reflection

Water and glass both have a greater index of refraction than air (light travels more slowly in them). In the case of a coin in a bowl of water (page 41) we saw that light emerging from water into air bends away from the perpendicular.

Suppose you are a diver with an underwater light, aiming it upward to the water surface. If you point it straight up, the light will go straight out into the air. If you aim it at a small angle to the surface, on the other hand, the refracted beam will bend away from the perpendicular as it goes into the air. The greater the angle of aim, the more the emerging beam will bend. You could aim it at such a large angle that the refracted beam would bend back into the water and not emerge into the air at all. This is total internal reflection and can occur when light goes from a medium of greater to a medium of lesser index of refraction. As a diver, you don't see much of the upper air world. There is a mirror effect when you look up as rays bounce back down from the surface.

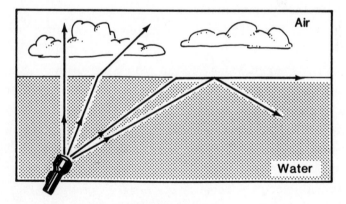

An underwater light aimed toward the surface refracts as it emerges into air. For a large-enough angle the refracted beam does not emerge at all but is "totally internally reflected."

In a long, thin glass rod total internal reflection occurs. Light entering one end of the rod makes a large angle with the surface of the glass and bounces back, crossing the rod and reflecting off the other side. As long as the glass doesn't absorb it, the light will zigzag its way down the rod, finally emerging from the far end. Rather than depend on a glass/air surface, with properties that change with dirt and scratches, manufacturers clad (coat) the glass strands with another material that has a lower index of refraction.

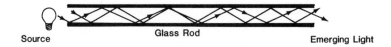
Source   Glass Rod   Emerging Light

The fact that light can be made to zigzag down a glass fiber is the principle of fiber optics. The fibers are flexible and can be wrapped together in bundles to make a cable both smaller and lighter than copper wire telephone cables.

Long-Distance Communication
To make use of glass fibers, engineers use semiconductor lasers designed to emit light at the wavelength to which the glass is most transparent. The laser beam from a tiny crystal, of course, has a wider spread than has the beam from a long gas laser tube. Focused by a lens, however, it fits well into a small fiber. Total internal reflection means that there is no loss through the walls. With little absorption by the glass, the signals travel many miles before they fade enough to need amplification.

As with electrical signals, optical signals gradually fade out. For long-distance conversations, there need to be repeaters, or boosters, along the lines to bring the signals up to their original strength and send them on their way. The new optical cables need fewer repeaters than wire cables do—which means that installing and maintaining them costs less.

Light in Fibers

Optical fibers are smaller, lighter, and cheaper than wire cables, and they carry more conversations. The cable shown in the photo on page 77 can carry 50,000 simultaneous conversations. Recent improvements have raised that number to several million and researchers are still working to improve the fibers and the lasers. Optical signals are not affected by electromagnetic interference, and it is harder to tap the lines. Clearly, optical fibers are going to replace long-distance wire cables throughout the country. The changeover began in 1983 with an AT&T optical telephone link between New York City and Washington, D.C. Shorter lines within some cities had been installed several years earlier; for example, the Walt Disney World EPCOT Center, a model city of the future, has fiber optic communications.

Tests are being made for an optical cable to be laid across the Atlantic Ocean in 1987 and possibly across the Pacific Ocean in 1988. It will be cheaper than another wire cable and will give better signals for international telephone calls than satellites.

In Chapter 4, we talked about digital recording of music and video. Telephone conversations can be similarly encoded into strings of 0's and 1's and translated at the other end into faithful reproductions of the voices. Since computer transmissions over telephone lines are by nature digital, the phone companies have chosen to use digital signals for all transmitting.

To get the greatest number of conversations into a fiber optic line, the 0's and 1's must follow each other at the fastest possible rate without overlapping. If you look at the drawing of light bouncing through a glass rod (page 74), you can see that various zigzag paths are possible for rays starting out at slightly different angles in the same rod. These different paths are called different modes. A multimode fiber permits many modes to travel simultaneously. Those that zigzag the most lag behind the others, and lagging modes cause a pulse of light to spread over a greater time. While that has little effect in a short line, over a long distance close-spaced pulses begin to overlap. Then the message is lost, just as a Morse code message would be lost if the dots and dashes piled on top of each other.

Manufacturers are now perfecting even finer fibers coated with material having an index of refraction that varies with distance out from the fiber. These fibers are designed to permit only a single mode to pass. Furthermore, since different wavelengths travel at different speeds in the glass, it is important to have only one wavelength. While people often refer to the single wavelength of a laser, generally there is a small range of wavelengths. For communications, engineers have now succeeded in making lasers that really emit only a single wavelength.

Short-Distance Communication

It is not only for fast, long-distance communications that fiber optic systems are useful. Short-range systems have been installed on some ships and airplanes where freedom from electrical interference is important. Fiber optic systems are going into automobiles to cut back on the excessive number of control wires. Some large businesses use them to link offices together or link computers within an office. Factories use them to carry signals to automated machinery. Doctors have flexible fiber optic instruments to look at your insides. Systems covering only a short distance can use the less expensive multimode fibers and tiny semiconductors which emit incoherent light. They are easier and cheaper to make than semiconductor lasers. Interestingly, scientists have recently learned that nature uses optical fibers to conduct sunlight down to the underground roots of plants.

Optical fibers may be preferred over electrical connections for even such short-distance applications as joining a computer to a printer. In fact, they are being considered for use inside a

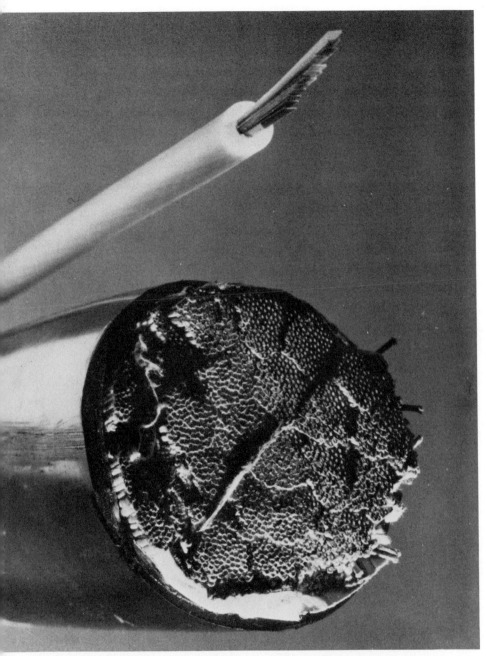

An ordinary 1200-pair wire cable looks large compared to the 144-fiber, half-inch-diameter lightguide cable. Yet the glass fibers can carry nearly 50,000 simultaneous conversations—more than triple the capacity of the larger cable.

computer to link one board with another, to link individual chips on a board, or even, on a microscopic scale, to link cells on a chip.

Optical links have the advantages of being faster and of requiring less power than systems that depend on moving electrons through wires. If electronic chips in a computer have optical inputs, they can match up directly with fibers coming in from medical sensors, from other computers, or from telecommunications. Optical fibers are not the only possibility, however, for optical links within a computer chip. Light might also pass through a short gap of free space or be conducted through thin transparent films. The technology of short-range optics is in an early stage of development.

Interferometry

In Chapter 1, we talked about the interference of two beams of light that occurs when one of them has traveled a slightly greater distance than the other and they are out of phase. There are instruments called interferometers that use this effect. One optical fiber is in a fixed location and the other is attached to a movable object. A tiny change in position, small compared to a wavelength of light (therefore less than about twenty millionths of an inch or fifty millionths of a centimeter), shows up as interference between the beams. Instruments which respond to the interference then move the object back slightly. Such instruments will be used in the control of large space platforms and in keeping many-segment telescope mirrors accurately in position.

Laser gyroscopes act like sensitive compasses for keeping ships and planes on course. A laser gyroscope depends on interference between two beams sent in opposite directions around a circular path. When the vehicle is turning in the direction of one beam, that beam will have a shorter path than the other,

Light in Fibers 79

and there will be interference when the beams are recombined. A set of three such instruments, at right angles to each other, can keep accurate track of an airplane's motions in three dimensions.

In fiber optic gyroscopes, split laser beams follow circular fibers in opposite directions. They can be made so small and rugged, and eventually will be so cheap, that even hang glider pilots will be able to use them.

There is an interesting use of lasers we haven't mentioned—the making of holographs. Let's look at that in some detail.

6
Reconstructed Light : Holography

One of the most fascinating uses of lasers is in making holograms. For a few years only laboratory scientists saw these three-dimensional views. Then holograms began to appear in museums, and finally in March 1984 the *National Geographic* magazine carried on its cover a holographic image of an eagle. Tens of millions of people saw that one, and a good many of them ran their fingers over it and then looked at the back side of the page trying to learn why that eagle looked so three-dimensional.

The eagle feels smooth to the touch and there is nothing on the back of the cover, but when you move your head from side to side, you get a different view from different angles. It is not at all like an ordinary flat photograph. You feel that you are looking at a small solid model of an eagle. Now there are similar holograms on greeting cards and record album covers. For these, as with the eagle hologram, you look at light reflected from them.

Museum displays generally use transmission holograms, meaning that you look at light coming through them. A transmission hologram is a piece of photographic film that has been exposed in laser light and then developed. For viewing, it is usually illuminated with the same laser. You do not have the feeling that you are seeing a picture on the surface, but rather that you are looking through the hologram to an apparent scene beyond. That scene actually was present while the exposure was being made, but it is no longer there when you are looking through the hologram. Nevertheless, it appears to be there. As you look through the film at different angles you see different views. By moving your head you can not only see that the objects have depth but can actually look around those in the foreground to see things behind them.

Although pictures from a stereo camera give a three-dimensional effect, it is from a single viewpoint. A hologram allows you to change the angle of view to the same extent that you could while looking through a window the same size as the film.

A very interesting hologram is one that records a printed page and a small magnifying glass over that page. When you look through the completed hologram you can move your head and see different letters magnified in turn, as though the page and the magnifying glass were still present.

That is a startling thing to see, and it's hard to believe that the lens is not still there. Even more strange is the fact that if you pick up the holographic film and examine it in room light, you see nothing on it. There is no picture such as the one on a photographic negative. What is a hologram anyway?

To put it briefly, a hologram is a photograph of an interference pattern, a pattern so fine that it can only be seen with a microscope. When laser light shines through or is reflected from the microscopic interference pattern, the original light waves that came from the scene are reconstructed and come to your eyes just as they would have from the scene itself.

A hologram is nothing at all like an ordinary photograph except that both store information on photographic film. To make a black-and-white photographic image, you focus light from a scene onto a film so that each point on the film corresponds to a single point on the scene. Information about brightness is stored at each point on the film. When you look at a print made from the processed negative, the pattern of light intensity reflected to your eyes from the print is similar to that coming from the original scene.

For a hologram, on the other hand, you do not make each point on a film correspond to a point on the scene. Instead, light from each point of the scene produces an interference pattern that spreads over the entire film. This gives information about the phase of the light waves from each point as well as the intensity of the light.

The hologram is not easy to understand because we rarely think in terms of interference patterns. In Chapter 1 we talked about interference of light and saw that if two coherent waves overlap on a screen, they make a pattern of light and dark as the waves reinforce or cancel each other. The coherent sources in that discussion were two pinholes, both illuminated by the same point source of light. The pinholes acted as two coherent point sources.

It takes two coherent light waves to form an interference pattern. Consider the geometrically simple case of a plane wave and a spherical wave. A plane wave has the same phase over a whole plane perpendicular to its direction of travel. Light from a distant source like the sun approximates a plane wave—or you can form a plane wave with a point source and a lens. A spherical wave, coming from a point source, has the same phase over an entire sphere. If a spherical wave and a plane wave, which are coherent, interfere on a film, they produce a particular pattern of dark and light concentric circles. Therefore, the hologram of a point source made with a plane-wave reference beam consists of such dark and light concentric circles.

Physicists are familiar with a very similar circular pattern which they call a zone plate. The black zones of a zone plate block light while the clear ones let light pass through. Scientists have long known that if they shine a plane light wave of the proper wavelength through a zone plate, the pattern will act like a converging lens and focus the light to a point, making a real image in front of the zone plate. (Simultaneously it acts as a diverging lens forming a virtual image behind the zone plate.)

The zone plate, consisting of a particular pattern of concentric circles. Alternate circular zones are blackened.

The significance of this is that a particular pattern results when you allow light from a single point to interfere with light from a plane reference beam. That pattern is exactly the one needed to focus the reference beam to a point. In other words, if you make a hologram of a single point and illuminate it with the reference beam, you'll get back a single point. That is true for any point or points.

Now think of a scene as a vast collection of points, all sending light toward the film. Each point produces, over the whole film, an interference pattern with the reference beam. The film records an overlapping maze of innumerable interference patterns. As a result, the developed film has an overall gray color to the naked eye. You cannot distinguish any detail.

However, when the film is illuminated by the reference beam, the complex interference pattern focuses the light to the whole series of original points. The light waves that come through the hologram are just like those that came from the original scene. It is because the original light waves have been reconstructed that you can see different views by moving your head, just as you could have with the original objects.

History

Optical holograms were invented almost by accident. In 1947 Dennis Gabor (1900-1979) was working in England with an electron microscope, an instrument that gives much higher magnification and can pick out much finer detail than an ordinary optical microscope.

In an optical microscope, the light passes through glass focusing lenses. In an electron microscope, electrons are focused by magnetic fields acting as lenses instead of by glass lenses. Gabor was dissatisfied with the quality of the images made by the magnetic lenses. However, he thought he could improve the images by making use of the wave properties of the electron. (Just as light has the properties of both particles and waves, so also do tiny atomic particles like electrons and protons have clear wave-like properties, including the ability to interfere with each other.)

Gabor decided to photograph the interference patterns he got with the electrons. That way he would get information about the phases as well as the intensities of the electron waves coming through his very thin samples. He hoped to use the photograph to re-form the waves and produce sharper images.

Gabor tried out the concept with light first, because light is easier to handle than electrons. Ideally it would be light of a single wavelength. To approximate this he used a mercury arc lamp—a lamp that emits all of its energy at a few discrete groups of wavelengths. He filtered out all but one bright narrow band of wavelengths which he allowed to shine through a clear photographic transparency with a few black letters on it. The letters served as objects. Light diffracted by the edges of the letters made interference patterns with the portion of the light that passed straight through the transparency. Each little part of each letter made its own interference pattern, and the hologram recorded all of the patterns.

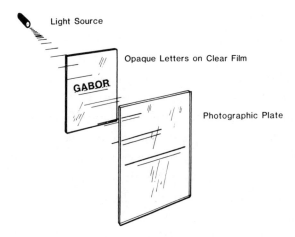

Dennis Gabor made the first hologram by shining light through a clear film with dark letters on it onto a photographic plate. An interference pattern, invisible to the eye, formed on the photographic plate.

Gabor removed the transparency with the letters on it. He put the developed hologram with its interference patterns back in place and let the same light shine through it. He could then see the images of the original letters just as though they were still there. That was the first hologram, although Gabor didn't call it that at the time. He wrote papers about "reconstructed wavefronts."

This technique did not give better electron micrographs. That's the way things go sometimes in research. However, the concept of optical holography was born; that is, interference patterns contain all the information necessary to reconstruct original wavefronts. In 1971, after the invention of the laser and further development of holography by other people, the Nobel committee awarded Dennis Gabor a Nobel Prize for his idea.

Improvements

There were some problems with Gabor's arrangement. For one thing, the sample had to be mostly transparent for the reference beam to pass straight through. Another point is that there are two images when a hologram is reconstructed. As with a zone plate, there is a real image where the rays come to a focus in front of the hologram. Also like the zone plate, when the interference bands of the completed hologram bend part of the illuminating beam to a focus in front of the hologram, they also bend another fraction of the beam so that it diverges. The diverging rays appear to come from a virtual image behind the hologram.

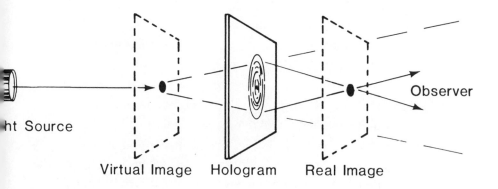

When Gabor's hologram was illuminated, its microscopic pattern of dark and light reconstructed the original light waves. He saw the original letters as though they were still there. In fact, he saw two images—one real and one virtual.

We talked about real and virtual images produced by lenses and mirrors in Chapter 2. Recall that it is a virtual image you see when you look at yourself in a plane mirror and also a virtual image you see when you use a magnifying glass over a printed page. Your eye gathers up the diverging rays and your brain interprets them as coming from an object beyond the lens or mirror. In the same way, you see a virtual three-dimensional image beyond the hologram. In Gabor's experiment, the two images were in line, the real one in front of the hologram and the virtual image behind it. That was confusing.

Emmet Leith and Juris Upatnieks at the University of Michigan later applied to holography an idea they had used with radar. (Radar is also a wave phenomenon and has some of the same features.) They aimed the reference beam at an angle to the object beam. That had the effect of spatially separating the real and virtual images. They were no longer in line. It also enabled researchers to use light reflected from an object. They could work with objects that weren't transparent.

In a typical arrangement, incoming light is divided into two parts by a beam splitter (a partially silvered mirror). The reference beam part is directed by mirrors to the film. The other part follows a different path and goes to the object. Some of the light reflects off the object and then goes to the film, where it makes interference patterns with the reference beam. When the object is removed and the developed hologram put back in place and illuminated with the reference beam, the real and virtual images are separate. It is the virtual image you see in looking through a transmission hologram and when looking at the hologram of the eagle on the cover of the *National Geographic*.

Laser Light for Holography

To get a good holographic image, it is necessary to have an accurate record of the interference pattern. That requires coherent light—that is, waves of the same wavelength and in phase.

Reconstructed Light: Holography

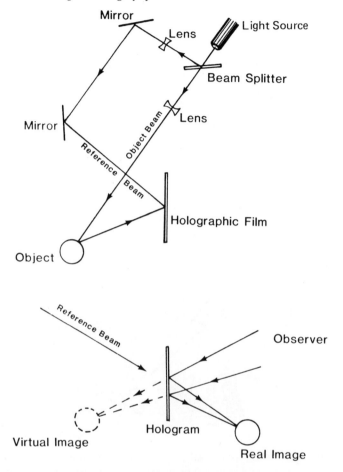

The reference and object beams come to the film from different angles to make the hologram. The lenses spread the beam to cover the object and film. Later when the hologram is illuminated with the reference beam, the real and virtual images are separated in space.

The mercury arc lamp was the best light source available at the time of the early holograms. However, it was necessary to filter out all but one spectral line. Also, as light waves from different parts of the arc were not in phase, it was necessary to use a pinhole in front of the arc to give essentially a point source. The net effect was to cut the light intensity greatly.

When they heard of Theodore Maiman's work with the laser, Leith and Upatnieks and other workers realized that the laser was ideal for holography. It was a source of intense light at a single wavelength, with its waves in phase. The waves would be in phase for a scene many inches across. Suddenly it became possible to make holograms of an entire tabletop of objects.

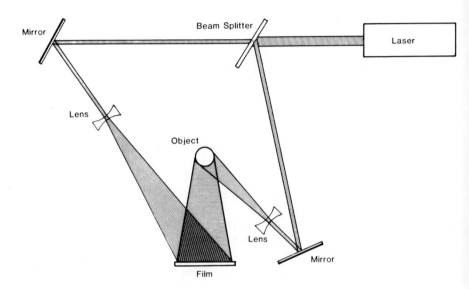

An arrangement for laser holography. Light waves reflected from the object and waves from the reference beam made interference patterns on the film.

Enthusiasm soared, and many people began making holograms. There were predictions of three-dimensional movies that did not need glasses for viewing. As it has turned out, however, holographic movies are not suitable for showing in theaters to large audiences. Movie makers in this country have no interest in productions to be seen by one or two people at a time, though a Russian group is making some.

Applications

What do people do with holograms? For one thing, they are used at the supermarket for checking out groceries. The checker doesn't need to align the product code carefully, but can run the package over the little window quickly. Inside the device is a spinning wheel containing a number of holograms. They act somewhat like lenses and scan laser light across the package at many angles, sending the reflected beam back to the detector. Usually one of the reflected beams gives the detector a pattern it can recognize.

Another use is in the continuing battle against credit card fraud. Cards now being distributed have holographic areas, which make them hard to counterfeit. Artists are intrigued with the hologram as a new art form, and there is even the Museum of Holography in New York.

One laboratory application of holography is in studies using a microscope. Minute animal forms in a sample of water have an annoying way of wiggling around while you are trying to focus on them. A brief, bright flash of laser light can "freeze" a holographic scene in an instant of time. The hologram reconstructs the original light waves, which will then pass through a microscope. You can focus up and down, getting a great depth of field, and the images of the little bugs hold still while you look at them.

For laboratory and industrial purposes, some of the most useful applications of the hologram are based on double exposures. The scientist makes a hologram of a device in normal condition. He then applies a stress and makes a second holographic exposure on top of the first. When reconstructed, slight changes in position of the material set up interference patterns between the two images recorded on the hologram. The photographs on pages 93 and 94 show "interferograms" for a variety of situations.

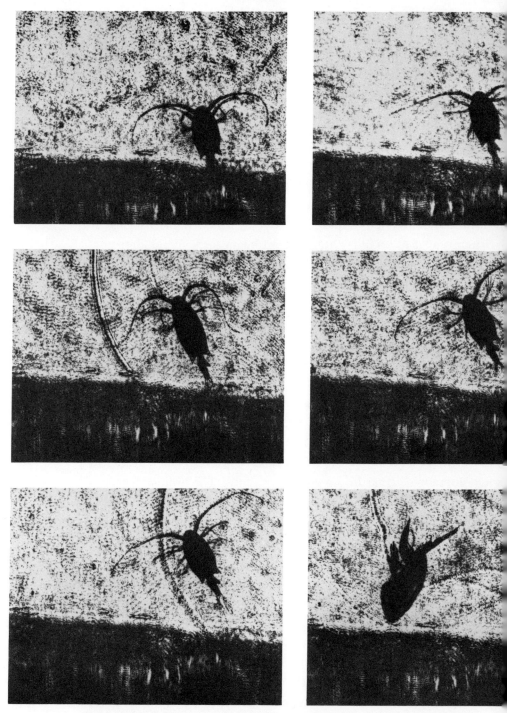

This sequence, about one-fiftieth of a second apart, is from a holographic movie of a tiny animal in plankton. The photographs, made through a microscope, allow a biologist to study how the little animal moves and feeds.

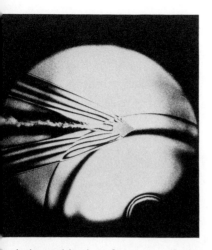

A holographic interferogram of the interaction of two shock waves. The spherical shock was from an electric spark; the other shock, from a high speed bullet.

A holographic interferogram of a light bulb. Argon gas is added to slow evaporation of the tungsten filament and prolong lamp life. The dark and light bands are due to change in density of the gas in the bulb.

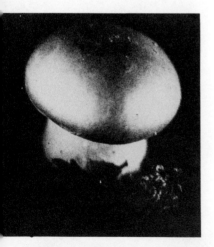 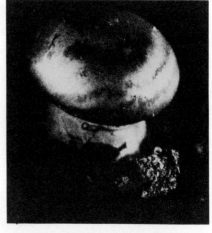

Left. A single-exposure hologram of a mushroom. *Right.* A reexposure after 25 seconds. The faint interference fringes on the right side of the mushroom show growth in that short period of time.

It is also possible to make holograms of systems that move back and forth in a regular manner—such as a vibrating radio speaker. Note in the photo that the vibration pattern is different at different frequencies.

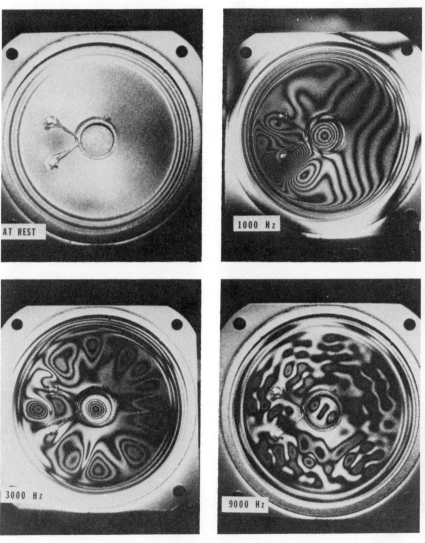

Holograms of a radio speaker vibrating at four frequencies: 0, 1000, 3000, and 9000 vibrations/second (hertz).

A Small Piece

There is another strange thing about a hologram: Every piece of it carries information for the whole scene. Take an ordinary photograph of a table top scene, cut the film in half and you have a picture of just half the scene. That is not so with a hologram. If you cut it and look through one half, you still see the entire scene.

The reason is that light from the entire scene reaches and makes interference patterns on every portion of the film. Every part of the hologram contains the pattern to reconstruct the original light waves. If you use a very tiny piece, the image will be dim and fuzzy, but it is surprising how small a hologram will give a recognizable reproduction. Some scientists suggest that memory in the human brain may be similar, with the same information stored globally throughout the brain instead of at a particular point.

The fact that so little space is necessary to record holograms means that they can be tightly packed. That makes them promising for data storage. Theoretically, the *Encyclopaedia Brittanica* would fit on a hologram the size of a piece of typing paper. Holograms might also store huge computer memories in a small space.

White Light Viewing

Though an expensive laser is necessary to make holograms, it would be preferable to view them with ordinary, inexpensive light. Stephen Benton of Polaroid Corporation devised one method of eliminating the viewing laser, a method which has been used for both transmission and reflection holograms. He makes a hologram in the usual way with laser light. Then he puts a narrow horizontal slit in front of the real image reconstructed from the hologram and, again with a laser, makes a new hologram of it through the slit.

That horizontal band of the hologram contains all the information needed to reconstruct the entire scene as seen through a narrow slit. (You lose the view if you move your head up or down.) However, you still have horizontal parallax—if you move your head from side to side, you still see around near objects. Since our eyes are side by side, it is horizontal parallax we generally use to judge distance.

A narrow slit bends, or diffracts, light by an amount that depends on the wavelength. The slit hologram will act much like a prism and spread out white light vertically into the spectral colors. If you raise and lower your head, you see the same scene in different colors. You probably noticed that effect in the eagle on the *National Geographic*. As explained in an article in the magazine, they used Dr. Benton's technique of making a second hologram through a narrow horizontal slit. The film was a special kind, called photoresist, which when developed, makes the interference pattern into a series of ultrafine ridges. Plastic films molded from the master copy were given a fine mirror coating of aluminum. The aluminum mirror reflects incident white light back through the interference pattern. This reflection technique was originated in 1962 by Y.N. Denisyuk of the Soviet Union. The result of the whole process is that you can look at the eagle in sunlight or ordinary room lights. As you move your head from side to side, you get the three-dimensional effect in one color. If you move your head vertically, you see the eagle in different colors.

So far, holography has been used primarily in military and industrial equipment and for artistic displays, but there are many ingenious people in other fields working with it. Many new devices are being developed. Someday soon you may find holograms used in your home—possibly even giving you three-dimensional television.

7
Invisible Light: Infrared

The term infrared conjures up thoughts of strange color photographs, burglar alarms, and spies in the dark. This invisible radiation seems very mysterious. What is infrared radiation? Where does it come from? How can we use it?

As Sir William Herschel discovered (page 23), infrared is electromagnetic radiation, just as visible light is. The chart in the first chapter shows the whole spectrum. Let's look at just the central portion.

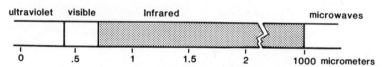

The infrared region of the electromagnetic spectrum ranges from 0.7 to about 1000 μm. It is far wider than the visible region, which is from 0.4 to 0.7 μm.

Infrared wavelengths are generally quoted in micrometers (millionths of a meter, μm), or in nanometers (billionths of a meter). Infrared wavelengths are longer than those of visible light. The frequencies, and therefore the infrared photon energies, are less than those of visible light. Because such low-energy photons don't activate the sensitive elements of the eye, we can't see radiation at infrared wavelengths. Starting at about 0.7 μm, it is necessary to use instruments to detect electromagnetic radiation.

The range of infrared photon energies is so great that several types of material are needed to transmit and detect different parts of the infrared spectrum. Scientists usually speak of near infrared, middle infrared, and far infrared regions.

The near infrared includes wavelengths from about 0.7 μm to about 1.5 μm. Sources of visible light generally emit moderate amounts of near infrared. Window materials that transmit visible light also transmit infrared wavelengths to 1.5 μm. Many detectors of visible wavelengths also respond to the near infrared region.

The term middle infrared is usually applied to wavelengths from 1.5 μm to about 10 μm. Most sources are weaker in this region than in the near infrared. Different materials transmit and detect middle infrared wavelengths, and it is often necessary to cool the detectors with liquid nitrogen to make them more sensitive.

The processes that produce infrared radiation are similar to those that produce visible light. In a gas, an electron falling from one atomic energy level to another generally emits a photon of visible light. If the energy difference between levels is very small, however, the photon will be at an infrared wavelength.

Atoms combined into gas molecules also have energy levels. These levels are usually close together so that, in a shift from one to another, an infrared photon is emitted. (That is the case for the carbon dioxide laser described in Chapter 3.)

In low-pressure gases, light or infrared energy is emitted at just a few isolated wavelengths. On the other hand, either a hot solid or the highly compressed gas of a star emits a continuum of electromagnetic radiation. This is because many close-spaced atoms are exerting forces on electrons. The electrons have a continuous range of energy levels to fall into and, therefore,

Invisible Light: Infrared

emit a continuous range of photon energies. The radiation may spread over ultraviolet, visible, and infrared wavelengths.

A good way to visualize a radiation source is to plot its intensity at each wavelength. Sunlight, for example, has a strong peak centered at about 0.5 μm, in the visible region of the spectrum. (It is interesting that our eyes have developed their greatest sensitivity just at the sun's peak output.) The intensity of sunlight falls off greatly in the infrared region, and even some of that is absorbed in our atmosphere.

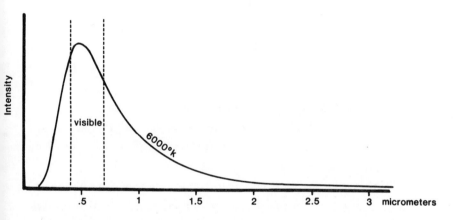

Intensity versus wavelength of radiation from an ideal emitter at the surface temperature of the sun.

The cooler a body is, the longer the wavelength of its peak emission. Radiation at human body temperature has its peak at around 10 μm. Furthermore, the total amount of energy emitted drops with decreasing temperature. The human body is a weak infrared emitter, and cooler objects put out even less.

You can see that researchers working at far infrared wavelengths deal with weak sources and need the most sensitive detectors. What's more, their measurements are confused by the fact that their equipment itself emits some of the same long wavelengths—just because of its temperature.

Infrared Photography

Imagine a store owner trying to protect his stock from burglary. He wants not only an alarm but also a picture of the thief in action—preferably taken without alerting the intruder that he has been photographed. This calls for infrared photography.

It is not as complicated as you might think. Kodak infrared film is available at camera stores. There are both black-and-white and color infrared film—even infrared movie film. The silver emulsion used in photographic film is, in its basic nature, sensitive to energetic blue photons. To enable you to record other colors, the color film you buy for your camera has been chemically treated to increase its sensitivity to the other wavelengths—green, yellow, and red. For infrared work, different chemicals sensitize the emulsion to wavelengths in the 0.7- to 0.9-μm range instead. (No chemicals have been found to extend the sensitivity beyond 0.9 μm.) The infrared film remains sensitive to blue light. If the source produces only infrared wavelengths, that is no problem. For daylight work, you need a filter to cut out blue.

With a few precautions, an ordinary camera can be made to work in the near-infrared region. Since the lens will focus the longer wavelengths at a slightly different position than it does visible light, it is necessary to increase the distance between the lens and the film a bit. The film speed is less than that of ordinary color film, and so longer exposures are required.

Flash and electronic flash emit in the 0.7- to 0.9-μm region as well as at visible wavelengths. With a filter over his flash to cut out visible light, the store owner can take a picture of an intruder in total darkness.

Camera stores carry Kodak booklets that give guidelines on focusing, exposures, filters, and other aspects of infrared photography. Even so, some experimenting is necessary to produce successful pictures.

Invisible Light: Infrared

Youthful burglars photographed by a hidden camera with infrared illumination.

©*Eastman Kodak Company*
Reprinted Courtesy of Eastman Kodak Company

Cloak-and-dagger work is not the only use for infrared photography. Naturalists can film nocturnal animals without frightening them. Aerial photography is clearer in infrared than in visible light because the long wavelengths are not scattered as much by atmospheric haze. It is even possible to take pictures in the "light" from hot electric irons.

Infrared reflection photograph made in the dark with the hot object emission from two electric irons (at the lower corners of the picture).

©*Eastman Kodak Company*
Reprinted Courtesy of Eastman Kodak Company

Daylight photography at infrared wavelengths gives some surprising false-color effects. Infrared color film makes green trees appear red, while normally red objects appear yellow or green. With black-and-white infrared film, the sky is gray and green trees are white.

A landscape with grass and trees so white that they appear covered with frost.

©*Eastman Kodak Company*
Reprinted Courtesy of Eastman Kodak Company

Infrared Detectors

There are two types of infrared detectors besides film: the "skin" type, which responds by absorbing radiant energy, and the "eye" type, which reacts to individual photons.

Detectors of the skin type are like sensitive thermometers. Some, called thermocouples, are electric circuits made of two different metals. Imagine a loop with, for example, copper wire partway around and wire of an alloy called constantan completing the loop. When one junction between the metals is held at a constant temperature in an ice bath and the other junction is warmed, a voltage develops between them. The voltage increases with the temperature difference making the thermocouple an electric thermometer. For infrared work it is calibrated to measure the amount of infrared energy absorbed in the junction. Detectors called thermistors are made of semiconductor materials. Their electrical resistance changes sharply with temperature. When calibrated, thermistors give a sensitive measure of absorbed radiation. A third "skin" type detector, the Golay cell, makes use of the expansion of a gas as it is heated by absorbed infrared energy. Thermocouples, thermistors, and Golay cells all depend on the heating effects of absorbed radiation. They react slowly.

The human eye is an example of the "eye" type. We are able to see because visible light photons strike sensitive molecules in the eye and produce changes that the brain interprets as color, brightness, and shape. Though infrared photons have energies too low to affect the molecules of the human eye, there are other materials that "see" infrared. Near-infrared photons can cause electron ejection (the same photoelectric effect that demonstrated the particle nature of light as described in Chapter 1). The collected electrons make an electrical current—its size measures infrared intensity.

For wavelengths longer than about 1.3 μm, photons have too little energy to kick electrons from any type of surface. In semiconductors which have been cooled, however, infrared photons can release lightly bound electrons so that they flow inside the material. Since photodetectors don't depend on heating anything, their response times are very short—maybe millionths of a second.

The world would look very different if human eyes responded to wavelengths near 10 μm instead of near 0.5 μm. We would see things not by reflected sunlight, since there is little sunlight at 10 μm, but by thermal radiation from the objects we are looking at. A group of things at the same temperature, like the different foods in a refrigerator, would all be equally dim. In comparison, a warm hand reaching into the refrigerator would glow brightly.

Certain snakes have, in addition to eyes, some "pit" organs which are sensitive to warmth. The pits do more than just sense heat the way your skin feels the glow from a fireplace. They send nerve impulses to the same part of the brain that the eyes do. Even in the dark the snake sees the warm body of a mouse.

Other Uses of Infrared Radiation

Scientists use infrared wavelengths for many purposes. Chemists use infrared to study molecules. Document analysts can make out faded details in ancient documents (such as the Dead Sea Scrolls) and in papers that have been charred by fire. Photographs on infrared film, taken from orbiting satellites, can detect diseased forests; the infrared appearance of sick leaves is different from that of healthy ones.

The human eye is not a good guide to infrared optical systems. Glass lenses and windows look clear to us because they pass visible light, yet they are useless for long infrared wavelengths since glass absorbs infrared radiation. Windows that pass infrared radiation are made of materials like sodium chloride (table salt) or the element germanium. Germanium is transparent to wavelengths longer than 2 μm but stops visible light. To your eye it looks like a dull gray metal and appears to be a most unlikely window or lens material.

Sodium chloride has good transmission properties for a range of infrared wavelengths but has an unpleasant way of fogging up if it absorbs moisture. The infrared engineer must choose materials which not only transmit long wavelengths but also survive the temperatures, pressures, shocks, and other conditions they will meet, for example, in space or on the battlefield.

Since the days of World War II, American soldiers have used devices called snooperscopes to detect the enemy in the dark. One type sends out its own infrared beam to reflect off objects. A second type picks up infrared emitted by warm bodies. In either case a lens forms an image on an infrared-sensitive photoelectric surface. The electrons emitted from that surface are accelerated by an electric voltage to a fluorescent screen. The device somewhat resembles a small television set, but gives a rather hazy view of the scene.

The devices that emit infrared of course provide targets for an enemy that has its own infrared detectors. On the other hand, the devices that pick up the very weak infrared emitted by warm bodies need much larger lenses, so are bigger systems that are more awkward for soldiers to carry.

Infrared detectors enable planes to make night landings. Missiles equipped with infrared detectors can home in on the hot engines of enemy aircraft. The detectors can also distinguish between warm, live enemy missiles and cool decoys flying along with them.

Hughes Aircraft Company makes a portable unit for civilian use called the Probeye. Firemen use it to find victims in smoky buildings and to locate hot spots behind walls. Policemen use it in the dark to find fugitives. Contractors use it to find heat leaks in buildings; industrial workers, to find electrical or frictional hot spots in equipment.

The man on the right is making an energy scan with the Probeye infrared viewer.

Infrared measurements are also used in medicine, mainly as a diagnostic tool. There are scans for breast tumors, scans of severely burned patients, and varicose vein studies. The doctor gets information about temperature, which is related to blood flow to the area. Abnormalities show up as temperature differences.

Infrared Astronomy

Our knowledge of the universe from studies in astronomy has leaped forward in recent years. New observations have given us new theories about pulsars, quasars, black holes, and the formation of stars and planets. Astronomers are extending their measurements from the visible to both longer and shorter wavelengths. Radio astronomy has given useful information, and there are now X-ray and gamma ray, as well as infrared, observations.

To avoid atmospheric absorption, infrared astronomy is now carried out on mountain tops. Even in the mountains, observations are often concentrated in the 8- to 14-μm region, where the atmosphere has a "window." Of course, observations from satellites avoid this problem.

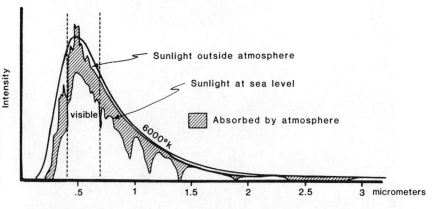

The intensity of sunlight as it reaches the earth's atmosphere has the curve shown. The shaded areas show atmospheric absorption. Transmission windows occur between bands of strong absorption. Though too low to be seen on the scale of this drawing, there is some infrared in sunlight extending to very long wavelengths, which can be measured in the 8- to 14-μm region.

The difficulty with infrared measurements is that the sources are very weak. Nearby infrared sources can confuse the readings. Because of their temperature, the telescopes and other equipment emit radiation at the same wavelengths as the objects being studied. The photodetectors themselves give off signals, since their electrons have heat energy. Strange signals arise from atmospheric temperature changes, passing clouds, and flying birds and insects. Even the sky itself emits long wavelength radiation.

To some extent astronomers overcome these problems by cooling the detectors with liquid nitrogen and using cooled "baffles" to help shield the instruments from extraneous radiation. Astronomers also use electronic tricks to subtract background radiation, leaving only the faint signal from the star of interest.

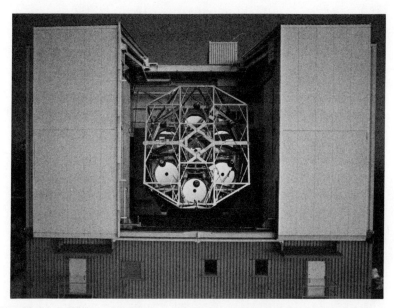

The Multiple Mirror Telescope in Arizona is designed for use in infrared, as well as visible, regions. It uses special techniques to reduce background signals.

Scientists studying our solar system use infrared measurements to get information about gases above the sun's surface and layers deep within the sun. Infrared measurements also tell them about the distribution of materials on the surface of the moon and about the ice molecules in Saturn's rings.

Astronomers studying the universe outside the solar system analyze infrared radiation from cool stars, which emit little visible light. They also study hot stars that are surrounded by great clouds of dust (the dust absorbs much of the visible radiation of the star and then emits infrared radiation). They believe that in some cases new stars are being born in the dust clouds. Recently there has been quite a stir because measurements from the infrared satellite hint that some other stars may have one or more planets.

Infrared measurements also may help to discover whether intelligent life exists anywhere else in the universe. We're sure to hear a great deal more about this as further infrared data are collected and analyzed.

8
The Future of Optics

The new optics has brought many changes into our lives and is sure to bring many more. Let's speculate about our future in a photonics world.

Cameras
Cameras today are better than ever. It used to be that lenses, ground on production lines, had to be of a spherical shape to keep costs reasonable. Now there are plastics of excellent optical properties that can be molded to any desired shape to overcome spherical and chromatic aberrations and give sharper images.

Engineers have developed the zoom lens in which multiple lens elements move back and forth relative to each other and to the film. As you turn a ring on the camera, you can keep an approaching skier in focus from the time he appears as a little dot on a mountain top until he is just a few feet from you and his grinning face fills the field of view.

This remarkable device would not be practical without newly developed coatings that greatly reduce the amount of reflection from an optical surface. Only a fraction of the light could get through the many glass parts of a zoom lens if it weren't for the antireflection coatings.

Optical sensors have made possible cameras with automatic exposure setting and automatic focusing. It is difficult to take a bad picture, and you almost wonder whether the camera of the future will tap you on the shoulder and point out a scene you should shoot. We can predict with certainty that many people will have video cameras and will show their own videos on their home television sets.

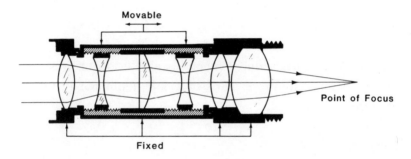

In the zoom lens, cams move groups of lens elements relative to other groups to focus the object on the film, whatever its distance.

Solar Energy

Huge numbers of photons from the sun shine on all parts of the earth each day. Architects are continuing to refine designs to use these photons in home heating. A significant number of solar homes have been built in the United States, even well north of the Sunbelt. The percentage of solar homes will continue to increase as older homes are replaced by solar designs.

Solar water heaters are proving economical in many climates, and scientists are still working to generate electricity from sunlight at a price competitive with power from coal and oil. Unless

The Future of Optics 113

unexpected success in the nuclear fusion program gives us unlimited cheap electric power in the near future, we may see orbiting satellites collecting sunlight twenty-four hours a day and beaming energy to earth in the form of microwaves or laser light.

New Uses of Lasers

Lasers, as we saw, are already being used in medical research and treatment. Some additional medical techniques now in the experimental stage will become commonplace. Lasers will be used to open clogged arteries and to vaporize tumors. In cancer treatment, doctors will inject into the body photosensitive dyes that are taken up specifically by cancer cells. Exposure to laser light will then kill those cells.

The use of lasers is becoming more common also in printing. You often see newspaper pictures labelled "laserphoto." The March 1984 *National Geographic* article on lasers described the use of lasers to transfer photos and maps to printing plates. In the future we can expect most printing companies to go to a fast system in which a laser first scans a page and stores the information in a computer. The computer then uses another laser to make printing plates.

When the laser at the checkout counter reflects off the universal product code on a package, the computer has to compare the image it receives with patterns stored in its memory. That is a form of pattern recognition. In the future, recognition of much more complicated patterns will be used to read handwriting on checks and to compare fingerprints. Pattern recognition will also enable "seeing" robots on assembly lines to pick up the correct bolt no matter how it might be lying in a box of parts.

Information and Communication

Lasers, combined with optical fibers, will give us a future crammed with information. Optical fibers can carry so many signals that we will have picture telephone communication in a few years. Some people predict that seeing the person at the other end of the telephone line will make telephoning so personal that businessmen will not need to travel to meet customers. They say that both long- and short-distance travel will be cut way back. Perhaps they are correct in that, but it is hard to imagine the computer and the video phone being used as a replacement for going to school or work.

As we saw earlier, digital music and digital video signals can be recorded as microscopic pits made on a disc by a laser. The same technique can be used to put gigantic memories into a small space. In the future, you may have an impressive home library on just a few discs. Researchers are now developing erasable optical memories for computers and other devices. While fixed memories are useful for libraries and archives, for some purposes you want to be able to change the data that have been stored. Japanese and American manufacturers are now coming out with optical memories which can be reused.

Electronic engineers can now make computers that respond to spoken commands instead of to typed instructions. However, a computer needs an enormous memory to recognize speech. You won't be able to talk to your computer until some kind of compact memory becomes cheap.

Futurists are looking to the time when there will be huge data banks on orbiting satellites, accessible to the whole world. Then we can all call up information from the world's great libraries, have access to experts on every subject, be able to look at all sorts of catalogs, and perhaps do all our shopping from our homes.

Satellites may also make life a lot safer. There are now navigation satellites that can give almost the exact position of a ship at sea. Satellites will soon keep track of trains and trucks on land as well. Further down the line, a person will be able to carry a pocket unit that communicates with an orbiting satellite to determine the individual's location within a few feet. That could mean quick help for lost hikers, for wandering children, and even for victims of kidnappers or muggers.

Finally, we may see the day when there are optical computers. An optical device that can replace a transistor in some applications has been developed. It is called a transphasor and it permits a small light signal to switch a much larger signal on and off.

Modern digital computers consist basically of vast numbers of interconnected transistors, tightly packed together. With the aid of electrical signals passing between them, they take in commands and data, do calculations, and store, or put out, commands and information. Eventually it may be possible to perform digital computing functions with optical, rather than electronic, components if engineers can overcome the many problems in manufacturing large numbers of closely packed optical components and of guiding light signals between them. This is much more difficult with light than with electrical signals.

There is much effort going into the development because optical components can potentially operate in a small fraction of the time required by electronic ones. There are some calculations so complex that they take too long even with today's fastest computers. With optical computers they might be possible.

Photonics is one of the most exciting areas of science now and for the future. Anyone interested in pursuing a career in science should seriously consider the varied and rewarding opportunities in this field.

Glossary

amplification—increase or enlargement.
angle of incidence—the angle which a ray of light, falling on a surface, makes with the perpendicular to that surface.
angle of reflection—the angle which a ray of light reflected from a surface makes with the perpendicular to that surface.
angle of refraction—when light passes at an oblique angle from one medium into another in which its velocity is different, it is bent. The angle of refraction is the angle in the second medium which the ray makes with the perpendicular.
astigmatism—one of several common defects of an optical lens or mirror, it causes rays from a single point of an object to fail to meet or appear to meet at a single focal point.
beam splitter—an optical device, such as a partially silvered mirror, that separates a beam of light into two beams that go different directions.
calibrate—to determine the graduations of a measuring instrument (such as the degree markings on a thermometer), or to correct those graduations by comparison with a standard.
center of curvature—the center of a circle which most closely fits a curve at the point of interest.
chromatic aberration—an effect in which different spectral colors of a light beam focus at different spots, resulting in colored fringes around an image in an optical instrument.
coherent light—light waves of the same wavelength and in phase.
concave—curving in, or hollow-shaped.
continuous spectrum—one spectral color shading into another with no gaps.

continuous wave (or c.w.)—this term describes a laser which emits a steady beam rather than bursts of light.
convex—arched up, or bulging out.
crest—the top of a wave.
detector—in this book, detector means an instrument for sensing light, infrared, or some other electromagnetic radiation.
diffraction—a phenomenon of waves in which each point on a wavefront acts as a source sending out a new wave in all directions. Hence light, water or other waves curve around obstacles.
dispersion—a phenomenon based on the fact that light of different frequencies has different velocities in a medium. This results in separation of colors in a prism, chromatic aberration in lenses, and smearing out of long-distance signals in fiber optic communication.
diverging—spreading apart.
electromagnetic theory—a mathematical description of the behavior of electromagnetic fields and their interaction with matter.
electromagnetic wave—a wave produced by the vibration of an electric charge.
electron—the most elementary charge of negative electricity; a tiny constituent of all atoms and the carrier of electric current in wires.
emission—flowing out.
energy level—a state of constant energy. Electrons in an atom are believed to be at several distinct energy levels. An electron remaining at one level does not radiate or absorb energy.
excited state—an energy level of an atom or molecule at a higher energy than the ground state.
far infrared—the longest infrared wavelengths, generally regarded as ranging from about 10 to about 1000 micrometers.
focus—a point at which light rays converge (or would converge if not interrupted), or a point from which rays appear to diverge.
Fraunhofer lines—dark lines in the sun's spectrum as observed from the earth. They are caused by absorption of sunlight in the cooler vapors of the outer layers of the sun, or in the earth's atmosphere.
frequency—the number of vibrations or cycles per unit time, usually per second.
ground state—the lowest stable energy level a system (such as an atom) can have.
gyroscope—a disc or wheel mounted to spin rapidly about an axis and free to rotate about one or both perpendicular axes. It tends to maintain its orientation in space and thus serves as a reference direction for navigation systems.
hertz—a unit of frequency equal to one cycle per second.

Glossary

hologram—a photograph of an interference pattern, which can be used to reconstruct the original wave.
image—the representation of an object produced by a lens, mirror or other optical system.
incoherent—not consistent. Incoherent light contains light waves of a variety of wavelengths, phases and directions.
index of refraction—The velocity of light in a vacuum divided by the velocity in a medium such as glass. It is different for different materials and also depends on the wavelength of the light.
infrared waves—invisible electromagnetic waves lying just beyond the red end of the visible spectrum; longer than visible waves, but shorter than microwaves and radiowaves.
interference—the mutual effect when two waves of any kind meet. It causes neutralization at some points, reinforcement at others.
interferometer—an instrument that uses interference to measure small lengths.
inverted image—a image that is turned upside down from the object.
ionize—to cause a neutral atom or molecule to gain or lose one or more electrons and thus become a charged particle. In this book, ionize has been used to mean removing an electron from an atom.
laser—an optical source which emits coherent light.
laser video disc—a disc, much like a phonograph record, which carries information as a series of microscopic spots burned by a high power laser. When read by a low-power laser, the information is converted to sight and sound by a television set.
lens—a piece of glass or other transparent material with two opposite smooth surfaces that are both curved, or one curved and the other flat.
metastable state—an energy state of an atom, higher than the ground state, that is relatively stable. The atom can remain in a metastable state much longer than in other excited states.
micrometer—one millionth (0.000001) of a meter.
middle infrared—usually refers to electromagnetic wavelengths from about 1.5 to about 10 micrometers.
mirror—any glass or smooth surface that forms images by reflection of light rays.
nanometer—one billionth (0.000000001) of a meter.
near infrared—electromagnetic wavelengths from the red end of the visible spectrum to about 1.5 micrometers.
nucleus—the central part of an atom containing most of the atom's mass and having a positive charge. The nucleus may consist of a single proton or a group of protons and neutrons.

parallax—the apparent displacement of an object as seen from two different points.
parallel—all parts equally distant—never meeting, however far extended.
pattern recognition—the process in which characteristics of a pattern are detected and manipulated with the aim of identifying a particular pattern. Human beings excel at recognizing patterns; some progress has been made in building machines that can do so.
perpendicular—at right angles.
phase— the position along a wave. If two waves have their maximum at the same time and place, they are said to be in phase. It is the phase of one wave relative to another that is the physically important quantity.
photoelectric emission—the ejection of electrons from a material surface due to light striking the surface.
photon—a light quantum or particle.
polarized light—light whose wave vibrations are restricted. For example, plane polarized light vibrates in a single plane.
population inversion—when most of the atoms in a material have their electrons at higher energy states, the atoms are said to have population inversion—in contrast to the normal situation with most atoms at low energy states.
prism—a transparent body with a triangular cross-section.
pumping—applied to lasers, the term means raising atoms to excited states by some method—such as optical pumping.
quantum—an elemental unit of some physical quantity such as energy or momentum.
quantum theory—a mathematical theory based on the idea that the emission and absorption of energy by atoms and molecules is not continuous, but takes place in steps.
real image—an image formed by light coming to points of focus.
reference beam—as applied to holography, a beam which goes directly to the film where it sets up interference patterns with light reflected from the object.
reflection—the return of waves after striking a surface.
refraction—the deflection from a straight path of a wave passing obliquely from one medium to another in which its velocity is different.
spectroscope—an optical instrument that separates light into its component colors and is used to study the wavelengths present in a light source.
spectrum—the spread of radiant energy arranged in order of wavelength.

Glossary

spherical aberration—an effect in which rays of light from one point of an object, reflected from different parts of a spherical mirror, focus at different spots—resulting in a blurred image.

stimulated emission—light emitted from an excited atom stimulated by the passage of another light photon.

thermistor—a device for measuring temperature based on a change in electrical resistance with temperature.

thermocouple—a device for measuring temperature based on the electric voltage it produces.

transparent—describes a material you can see through.

transphasor—an optical device which behaves much like an electronic transistor.

trough of a wave—the bottom portion of the wave.

ultraviolet—invisible electromagnetic waves lying just beyond the violet end of the visible spectrum and of shorter wavelength.

universal product code—the identifying pattern of dark lines found on packages at the market.

virtual image—an apparent image at a location that light does not actually reach—as the image in a plane mirror.

wavelength—the distance between successive crests, troughs or other identical parts of a wave.

Further Reading

BOOKS

Beesley, M. *Lasers and Their Applications.* London: Taylor and Francis, 1972.

Bova, Ben. *The Amazing Laser.* Philadelphia: Westminster Press, 1971.

Hecht, Jeff, and Teresi, Dick. *Laser: Supertool of the 1980s,* New Haven: Tichnor and Fields, 1982.

Kettelkamp, Larry. *Lasers, the Miracle Light.* New York: William Morrow, 1979.

Klein, H. Arthur. *Holography.* Philadelphia: Lippincott, 1970.

Kock, Winston E. *Lasers and Holography.* 2nd edition, New York: Dover, 1981.

Schneider, Herman. *Laser Light.* New York: McGraw-Hill, 1978.

Stambler, Irwin. *Revolution in Light: Lasers and Holography.* New York: Doubleday, 1972.

Wenyon, Michael. *Understanding Holography.* New York: Arco, 1978.

PERIODICALS

Boraiko, Allen A. "Lasers—A Splendid Light." *National Geographic*, Vol. 165, March 1984, pp. 335-363.
Bova, Ben. "A Laser Lift." *Science 84*, Vol. 5, September 1984, pp. 78-80.
Caulfield, H. John. "The Wonder of Holography." *National Geographic*, Vol. 165, March 1984, pp.364-377.
Eberhart, J. "A Cold Eye on the Hot Sky." *Science News*, Vol. 123, January 15, 1983, pp. 36-37.
Edelson, Ed. "Holography—out of the lab at last." *Popular Science*, Vol. 224, March 1984, pp. 90-94.
Fisher, Arthur. "Medical Photography." *Popular Science*, Vol. 223, November 1983, pp. 16-17.
Free, John. "Heatless Laser Etching." *Popular Science*, Vol. 223, December 1983, p. 114.
"The Global Telecommunications Revolution." *Science Digest*, Vol. 92, March 1984, pp. 45-50.
Gold, Michael. "The Cosmos Through Infrared Eyes." *Science 84*, Vol. 5, March 1984, p. 16.
Schefter, Jim. "Orbiting Telescope Maps an Infrared Universe." *Popular Science*, Vol. 223, December 1983, pp. 88-92.
"Sharpened Bits." *Scientific American*, Vol. 250, February 1984, pp. 72-73.
Simon, Cheryl. "IR Restores Vision to Blinded Pilots." *Science News*, Vol. 124, December 24/31, 1983, p. 413.

Index

A
aberration
  chromatic, 43
  spherical, 37-38
antireflection coating, 111-112
Apollo vehicle, 30, 57
astronomy, 14, 107-109
atom, pictorial representation, 48

B
beam splitter, 88-90
Bell, Alexander Graham, 71
Benton, Stephen, 95-96

C
coherent light, 47, 50-51, 53
color
  film, 100
  and photon energy, 27-28
  and wavelength, 23
  in white light, 21-22
computer data storage, 61

D
Denisyuk, Y.N., 96
diffraction, 20, 21, 38, 96
digital
  communication, 75
  encoding, 61
  music, 61
distance measurement, 57, 66

E
Einstein, Albert, 27, 52-53
electromagnetic
  continuum, 98
  spectrum, 23-25, 31, 98
  wave, 21
electron, 47-50, 52, 54
  microscope, 85
energy level of atom, 48-49, 52-53
EPCOT Center, 75
eye, human, 17, 43, 97, 99, 104
eyeglasses, 43

F

fiber optics, 71-80
  communication, 14
fibers
  glass, 74-76
  multimode, 75-76
  single mode, 76
flashlamp, 54-55
Fraunhofer lines, 45

G

Gabor, Dennis, 85, 87
glass fibers. *See* fibers.
Golay cell, 103-104
ground state, 48, 50, 52, 54
gyroscope
  fiber optic, 79
  laser, 78

H

helium in sun, 45
Herschel, Sir William, 23-24, 97
holograms, 81-96. *See also* lasers.
  interference in, 82-86, 95
  uses, 91-95
  white light viewing, 95-96
Huygens, Christian, 18

I

infrared. *See also* spectrum, radiation.
  astronomy, 107-109
  detector, 103-104
  laser, 61-63, 64, 65, 98
  medical uses of, 107
  photography, 100-102
  window materials, 105
  uses, 105-109
interferometry, 78-79, 91-94

L

laser, 41-70
  argon, 60-61
  carbon dioxide, 61-63, 64, 65, 98
  chemical, 62, 64-65
  continuous wave, 59
  drilling, 62
  glass, 68
  helium-neon, 59-60
  in holography, 88-90
  infrared, 61-63, 64, 65, 98
  liquid, 65-66
  medical uses of, 60, 62, 63, 66, 113
  organic dye, 66
  printing, 113
  pulsed, 59
  pumping of, 59, 69
  ruby, 54-57, 59, 66-67
  safety precautions, 70
  semiconductor, 68-70, 72, 74
  tunable, 66
  versus other sources, 47
  video discs, 61
  weapons, 64-65
  welding, 62
  YAG, 66
Leith, Emmet, 88, 90
lenses, 38, 41-43
  camera, 111-112
  zoom, 111-112
light
  amplification of, 54-56
  coherent, 47, 50-51, 53
  dual nature of, 28, 31
  interference of, 18, 20-21
  particle theory of, 17, 28, 48
  theory of, 18, 21, 25, 27-28, 48
  show, 61
  sources, 47
  speed of, 28-29
light-year, 30

M

magnifying glass, 42

Index

Maiman, Theodore, 54-57, 59, 90
Maxwell, James Clerk, 27
metastable state, 54, 66
Michelson, Albert, 29
mirror
  curved, 34-38
  focus of, 34-36, 38
  parabolic, 37-38
  plane, 32-33
multiple mirror telescope, 108

N
National Geographic, 81, 88, 96, 113
navigation satellite, 114
Newton, Sir Isaac, 17, 18, 21, 28, 48
nuclear fusion, 68, 69

O
optical
  computer, 115
  fibers, 72, 74-77
  memory, 114
  processor, 14
  transistor, 115
orbiting data bank, 114

P
parabolic mirror, 37-38
parallel light rays, 34, 35, 37-38, 42
particle theory of light, 17, 28, 48
photodetector, 104
photoelectric emission, 27
photon
  absorption, 48-49, 52-54
  emission, 50, 52-56, 69
  energy, 27-28, 48-50, 52-54, 97, 98, 104
  unit of light, 13, 27-28, 31, 48-50, 52-54
photonics, definition of, 13-15
photophone, 71
polarization of light, 25-27

Polaroid material, 25-27
population inversion, 54, 57, 69
prism, 21-23, 44, 66
Probeye, 106

Q
quantum theory, 48

R
radiation
  human body, 99
  infrared, 97-109
radio waves, 24-25, 30, 51
ray diagrams, 33-38, 42
real image, 35, 42, 84, 87-88
reference beam, 83, 84, 85, 87, 88, 89
reflection
  diffuse, 33
  Huygens' theory, 18
  Newton's explanation, 17
  parabolic mirror, 37-38
  plane mirror, 32-33
  spherical mirror, 33-37
refraction
  by lenses, 38, 41-43
  at plane boundary, 39-41
  Newton's explanation of, 17
  wave theory, 18
repeaters, telephone, 74
resonant cavity, 57
Roemer, Olaf, 28-29

S
satellite
  data bank, 114
  navigation, 114
Schawlow, Arthur, 54
smart bombs, 65
snake's pit organ, 104
snooperscope, 105
solar energy, 112-113
spectroscope, 44-45

spectrum
 absorption, 45
 continuous, 44, 45, 65-66
 electromagnetic, 23-25, 31, 98
 infrared, 23, 97
 of mercury, 44
 of sunlight, 21, 23
stimulated emission, 52-56, 69
sunlight intensity, 107

T
telescope, 13, 14, 28-29, 37-38,
 45, 108
thermistor, 103
thermocouple, 103
total internal reflection, 73-74
Townes, Charles, 54
transatlantic cable, 75
transphasor, 115

U
underwater laser signals, 61
universal product code, 60, 91, 113
Upatnieks, Juris, 88, 90

V
virtual image, 36, 42, 84, 87-88

W
wave
 frequency, 23
 interference of, 18-19
 one-dimensional, 18
 speed, 23
 theory of light, 18, 21, 25,
  27-28, 48
 two-dimensional, 19

Y
Young, Thomas, 18, 20-21

Z
zone plate, 84, 87